Research Administration and Technology Transfer

James T. Kenny, *Editor*
Auburn University, Montgomery

NEW DIRECTIONS FOR HIGHER EDUCATION
MARTIN KRAMER, *Editor-in-Chief*
University of California, Berkeley

Number 63, Fall 1988

Paperback sourcebooks in
The Jossey-Bass Higher Education Series

Jossey-Bass Inc., Publishers
San Francisco • London

James T. Kenny (ed.).
Research Administration and Technology Transfer.
New Directions for Higher Education, no. 63.
Volume XVI, number 3.
San Francisco: Jossey-Bass, 1988.

New Directions for Higher Education
Martin Kramer, *Editor-in-Chief*

New Directions for Higher Education is published quarterly
by Jossey-Bass Inc., Publishers (publication number USPS
990-880). *New Directions* is numbered sequentially—please
order extra copies by sequential number. The volume and issue
numbers above are included for the convenience of libraries.
Second-class postage paid at San Francisco, California, and at
additional mailing offices. POSTMASTER: Send address changes
to Jossey-Bass Inc., Publishers, 350 Sansome Street, San Francisco,
California 94104.

Editorial correspondence should be sent to the Editor-in-Chief,
Martin Kramer, 2807 Shasta Road, Berkeley, California 94708.

Library of Congress Catalog Card Number LC 85-644752
International Standard Serial Number ISSN 0271-0560
International Standard Book Number ISBN 1-55542-886-X

Cover art by WILLI BAUM

Manufactured in the United States of America. Printed on acid-free paper.

Ordering Information

The paperback sourcebooks listed below are published quarterly and can be ordered either by subscription or single copy.

Subscriptions cost $48.00 per year for institutions, agencies, and libraries. Individuals can subscribe at the special rate of $36.00 per year *if payment is by personal check.* (Note that the full rate of $48.00 applies if payment is by institutional check, even if the subscription is designated for an individual.) Standing orders are accepted.

Single copies are available at $11.95 when payment accompanies order. (California, New Jersey, New York, and Washington, D.C., residents please include appropriate sales tax.) For billed orders, cost per copy is $11.95 plus postage and handling.

Substantial discounts are offered to organizations and individuals wishing to purchase bulk quantities of Jossey-Bass sourcebooks. Please inquire.

Please note that these prices are for the calendar year 1988 and are subject to change without notice. Also, some titles may be out of print and therefore not available for sale.

To ensure correct and prompt delivery, all orders must give either the *name of an individual* or an *official purchase order number.* Please submit your order as follows:

Subscriptions: specify series and year subscription is to begin.
Single Copies: specify sourcebook code (such as, HE1) and first two words of title.

Mail orders for United States and Possessions, Australia, New Zealand, Canada, Latin America, and Japan to:
Jossey-Bass Inc., Publishers
350 Sansome Street
San Francisco, California 94104

Mail orders for all other parts of the world to:
Jossey-Bass Limited
28 Banner Street
London EC1Y 8QE

New Directions for Higher Education Series
Martin Kramer, *Editor-in-Chief*

HE1   *Facilitating Faculty Development,* Mervin Freedman
HE2   *Strategies for Budgeting,* George Kaludis
HE3   *Services for Students,* Joseph Katz

HE4   *Evaluating Learning and Teaching,* C. Robert Pace
HE5   *Encountering the Unionized University,* Jack H. Schuster
HE6   *Implementing Field Experience Education,* John Duley
HE7   *Avoiding Conflict in Faculty Personnel Practices,* Richard Peairs
HE8   *Improving Statewide Planning,* James L. Wattenbarger, Louis W. Bender
HE9   *Planning the Future of the Undergraduate College,* Donald G. Trites
HE10   *Individualizing Education by Learning Contracts,* Neal R. Berte
HE11   *Meeting Women's New Educational Needs,* Clare Rose
HE12   *Strategies for Significant Survival,* Clifford T. Stewart, Thomas R. Harvey
HE13   *Promoting Consumer Protection for Students,* Joan S. Stark
HE14   *Expanding Recurrent and Nonformal Education,* David Harman
HE15   *A Comprehensive Approach to Institutional Development,*
        William Bergquist, William Shoemaker
HE16   *Improving Educational Outcomes,* Oscar Lenning
HE17   *Renewing and Evaluating Teaching,* John A. Centra
HE18   *Redefining Service, Research, and Teaching,* Warren Bryan Martin
HE19   *Managing Turbulence and Change,* John D. Millett
HE20   *Increasing Basic Skills by Developmental Studies,* John E. Roueche
HE21   *Marketing Higher Education,* David W. Barton, Jr.
HE22   *Developing and Evaluating Administrative Leadership,* Charles F. Fisher
HE23   *Admitting and Assisting Students After Bakke,* Alexander W. Astin,
        Bruce Fuller, Kenneth C. Green
HE24   *Institutional Renewal Through the Improvement of Teaching,*
        Jerry G. Gaff
HE25   *Assuring Access for the Handicapped,* Martha Ross Redden
HE26   *Assessing Financial Health,* Carol Frances, Sharon L. Coldren
HE27   *Building Bridges to the Public,* Louis T. Benezet, Frances W. Magnusson
HE28   *Preparing for the New Decade,* Larry W. Jones, Franz A. Nowotny
HE29   *Educating Learners of All Ages,* Elinor Greenberg, Kathleen M.
        O'Donnell, William Bergquist
HE30   *Managing Facilities More Effectively,* Harvey H. Kaiser
HE31   *Rethinking College Responsibilities for Values,* Mary Louise McBee
HE32   *Resolving Conflict in Higher Education,* Jane E. McCarthy
HE33   *Professional Ethics in University Administration,*
        Ronald H. Stein, M. Carlota Baca
HE34   *New Approaches to Energy Conservation,* Sidney G. Tickton
HE35   *Management Science Applications to Academic Administration,*
        James A. Wilson
HE36   *Academic Leaders as Managers,* Robert H. Atwell, Madeleine F. Green
HE37   *Designing Academic Program Reviews,* Richard F. Wilson
HE38   *Successful Responses to Financial Difficulties,* Carol Frances
HE39   *Priorities for Academic Libraries,* Thomas J. Galvin, Beverly P. Lynch
HE40   *Meeting Student Aid Needs in a Period of Retrenchment,* Martin Kramer
HE41   *Issues in Faculty Personnel Policies,* Jon W. Fuller
HE42   *Management Techniques for Small and Specialized Institutions,*
        Andrew J. Falender, John C. Merson
HE43   *Meeting the New Demand for Standards,* Jonathan R. Warren
HE44   *The Expanding Role of Telecommunications in Higher Education,*
        Pamela J. Tate, Marilyn Kressel
HE45   *Women in Higher Education Administration,* Adrian Tinsley,
        Cynthia Secor, Sheila Kaplan

HE46   *Keeping Graduate Programs Responsive to National Needs,*
       Michael J. Pelczar, Jr., Lewis C. Solomon
HE47   *Leadership Roles of Chief Academic Officers,* David G. Brown
HE48   *Financial Incentives for Academic Quality,* John Folger
HE49   *Leadership and Institutional Renewal,* Ralph M. Davis
HE50   *Applying Corporate Management Strategies,* Roger J. Flecher
HE51   *Incentive for Faculty Vitality,* Roger G. Baldwin
HE52   *Making the Budget Process Work,* David J. Berg, Gerald M. Skogley
HE53   *Managing College Enrollments,* Don Hossler
HE54   *Institutional Revival: Case Histories,* Douglas W. Steeples
HE55   *Crisis Management in Higher Education,* Hal Hoverland, Pat McInturff,
       C. E. Tapie Rohm, Jr.
HE56   *Managing Programs for Learning Outside the Classroom,*
       Patricia Senn Breivik
HE57   *Creating Career Programs in a Liberal Arts Context,* Mary Ann F. Rehnke
HE58   *Financing Higher Education: Strategies After Tax Reform,*
       Richard E. Anderson, Joel W. Meyerson
HE59   *Student Outcomes Assessment: What Institutions Stand to Gain,*
       Diane F. Halpern
HE60   *Increasing Retention: Academic and Student Affairs Administrators
       in Partnership,* Martha McGinty Stodt, William M. Klepper
HE61   *Leaders on Leadership: The College Presidency,* James L. Fisher,
       Martha W. Tack
HE62   *Making Computers Work for Administrators,* Kenneth C. Green,
       Steven W. Gilbert

Contents

Editor's Notes 1
James T. Kenny

1. Global Technology Diffusion and the American 5
Research University
Thomas C. Collins, Sheadrick A. Tillman IV
Intense global technological and market competition is reshaping the
world economic and political order in ways that have important long-term
implications for the organization of research in America's universities.

2. The New Frontier of Technology Transfer 21
Gerald A. Erickson, Donald R. Baldwin
Research results that are increasingly being converted into commercial
products, processes, or services covered by patents or license arrangements
herald a problematic yet promising new era in higher education.

3. Higher Education in Corporate Readaptation 37
Ronald R. Sims
As America's businesses restructure and streamline operations to meet new
foreign and domestic challenges, colleges and universities are offering help
in the form of flexible informational, technical, and instructional services.

4. The University in Service to State and Local Government 49
James T. Kenny
In a rapidly expanding service environment, governmental entities are
looking to campus-based applied research as they seek to modernize and
improve public services.

5. Managing a Modern University Research Center 61
John G. Veres III
Some specialized university research units see exciting opportunities in
an environment in flux and are responding imaginatively and effectively
through the application of new and somewhat unconventional management principles.

6. The Future of University Research Administration 73
Mark Elder
Computerization, a new mood of interinstitutional cooperation, and the
emerging triad of university, business, and government will create major
changes in what universities will have to manage in the research enterprise and how they will manage what they have.

Index 87

Editor's Notes

The management of university research presents major challenges for modern academic planners and administrators. America's need for new technology and knowledge to ensure its market competitiveness places new strains on an already taxed educational system. At a time when the nation's colleges and universities must contend with problems related to aging and underfunded research facilities, the public and private need for a bountiful and problem-focused science seems greatest. In a fiscal mood of austerity and cost containment in higher education, our national economy, dependent as it is on scientific discovery and information transfer, appears troubled. A $170 billion trade deficit dims the national outlook, reflecting serious setbacks in manufacturing and business's export capacity. In this environment, national science policy seems to lack coherence and direction. The government appears to be sluggish in its support of basic research and unable to provide convincing leadership in the promotion of much-needed technology transfer.

Compounding the dilemma of declining productivity is the knowledge that America lacks a work force that is sufficiently trained to meet the expanding technological requirements of a modern society. The human resource problem touches universities at many points and is central to the strained dialogue on education's responsibility and its future. While the barriers to improved economic performance may not be insurmountable, they are real and, in the aggregate, determine the conditions under which today's research administrators and campus leaders must act.

University research and technical services, and the management of these entities, cannot be understood apart from the relationship of higher education to society, its difficulties, and its long-term aspirations. American universities, in spite of their internal preoccupations, occupy a unique position in today's information-driven world. Increasingly, they are turned to and asked to take on a new, responsible relevance. In the main, it is argued that if our nation is to prosper, universities must sustain their involvement in national problem solving. It is also posited that if higher education is to ensure for itself stability, vitality, and a preferential position in the society of tomorrow, some important normative and operational changes will need to be considered.

Our institutions of learning share—and, it is hoped, will continue to share—a special responsibility in the discovery and generation of new knowledge. Traditionally, the fruits of investigation have been offered to society in the form of research findings, field investigations, technical

1

assistance, publications, and training. Universities must continue to provide these. At the same time, enabling conditions for the successful continuation of the research enterprise must be guaranteed. From the administrative point of view, the nurturing of specialized units, ensuring their continued financial support and fostering a spirit of free exchange, are essential supervisory objectives. Effective leadership has nearly always involved the selection of outstanding personnel and the securing of first-class research and program facilities. Moreover, administrators must create a milieu where innovation and productivity can be identified and rewarded consistently. In today's world, however, those who manage or guide must be cognizant of and responsive to accelerated change and the external configuration of forces, trends, and opportunities that make up the larger global system.

The current generation of university leaders must understand that nearly as important as the need to generate new knowledge is the need to forge successful institutional links that will quickly effect the transfer of such knowledge. Improvements must now be made to increase the flow of data and technical assistance from the laboratory to the marketplace, from the scientist to the worker, and from the campus to the arena of public policy. At every turn, one sees a growing and nearly insatiable need for information and assistance that are timely and deliverable on demand.

Persistent pressures for modification are also felt in instruction. Whom and what colleges teach is an important and frequently debated issue. An earnest search for balance goes on, as campuses try to provide a high-quality, sound educational experience for students, which is also attuned to the world of work that students must encounter. Even so, the most salient pressures are felt in campus research and technical services, areas whose programs, by their very nature, must be more immediately responsive to the external pulse.

In the service environment, federal agencies—and, increasingly, state and local units of governance—seek to upgrade their administrative skill levels and delivery modes to cope with an increased public demand for programs, regulatory participation, and problem solving. The polity's woes are real and growing, and campus units are being called on to provide effective solutions to deep-rooted problems. At the same time, the corporate world is undergoing large-scale restructuring, is streamlining operations, and is seeking new efficiencies to meet the challenge of intense competition in the evolving world economic order. The promise of increased productivity, through the widespread adoption of technology and the prompt application of new knowledge, is taken as an article of faith by today's business leaders. Therefore, in the marketplace and in the society beyond the ivy-covered quadrangles, there exist some splendid new opportunities for research and education.

Some institutions of higher education are cognizant of their special responsibilities concerning information exchange. Research parks, business incubators, and innovation centers now dot the countryside and nestle, although at times uncomfortably, in the groves of academe. Some thinkers have predicted the emergence of a progressive and unparalleled triad of university, business, and government. One might hope that our centers of learning could act as catalysts for, or elements in, some positive dialectical evolution. History offers no certainties and little guidance on this point. We are admonished by Arnold Toynbee, however, that "the nemesis of creativity is the idolization of ephemeral technique." Perhaps, then, academic administrators should critically examine the organizational bases and current strategies of research and service management. At the very least, institutions, while retaining their basic identities, should perhaps become more responsive to external opportunities. Academicians must find the consensus and will to adapt their services to the requirements of a restructured economic milieu. They may have to become somewhat more aggressive in the representation of their products, more effective in data diffusion, and more alert to new opportunities for service and meaning in a demanding, quick-paced, and informationally appetitive environment. Leaders would also do well to remember that the day-to-day problems encountered by their research and service teams tend to be fiscal, contractual, and legal; this is to say that such problems are substantially different from the more predictable concerns of academic departments. In an increasingly litigious society, those who would introduce new products and services must be cognizant of issues related to human and animal subjects in research, hazardous substances, product and service liability, and a growing body of regulations. In today's environment, one sees a greater emphasis on risk assessment, risk management, and other monitoring activities. Administrators must therefore realize that as university services expand, so also must the infrastructure of related institutional policies.

Some institutions, even among those that are well equipped and seemingly well positioned, are not seeking broad internal agreement or forging the consensus that will give direction to their role in information exchange. Such institutions must make strenuous efforts to examine their educational values and to consider the future of their research and service missions. In this connection, they should also review their link with external institutions and the purposes of those links.

This volume, in practical terms, seeks some answers to a central question: In view of the enormous systemic changes we face as a nation and as a civilization, what can a modern university do or prepare to do to enhance its own capacity to provide abundant, valuable information and relevant external programs? In seeking useful answers, one cannot ignore the voices of campus units and administrators who work on the cutting edge of research and services. Not only should they be heard, but special efforts should also be made to bring their enthusiasm, ideas, and concerns into an expanded dialogue on what higher education is today and where it must go.

The chapters in this sourcebook offer analyses of some major changes taking place in today's university environment. Each of the contributors, representing various fields, views the improvement of information exchange, and organizing to meet that challenge, as a core concern of modern research management. The contributors were asked to share their views as practitioners and to offer their diverse and interesting perspectives on the changes in management brought about by global technology diffusion, national science policy, technology transfer, university entrepreneurship, corporate restructuring, governmental service requirements, and the evolution of technical service programs. In providing their insights, the contributors suggest strategies and policy orientations that should provoke thoughtful reflection as today's campus leaders consider creative adjustments to the demands of tomorrow.

James T. Kenny
Editor

James T. Kenny is vice-chancellor for research and development at Auburn University, Montgomery. He chairs that institution's Research Council and has worked in the organization of cooperative university, corporate, and governmental programs.

A changing configuration of world political and economic power has long-term implications for American universities and the organization of university-based research.

Global Technology Diffusion and the American Research University

Thomas C. Collins, Sheadrick A. Tillman IV

Rostow (1987), looking at historical and current trends in Soviet-American relations, predicted the demise of global political bipolarity and the likely end of the Cold War as we know it. His well-received article touched on points that have distinct implications for the future of American universities and the research enterprise. Rostow, who served as Director of Policy Planning at the U.S. Department of State in the early 1960s and later as a national security adviser to President Johnson, said, "The possibility of a soft landing for the Cold War is strengthened by two related revolutions that have been proceeding concurrently over the past decade. One is a major technological revolution generated in the advanced industrial countries, the other an educational revolution mounted in the more advanced developing countries, which is putting them in a position to absorb and apply the new technologies" (p. 840).

Before looking at the responsibilities and likely role of tomorrow's research university in view of these developments, let us examine some trends that have historically given definition to higher education's relationship to society. In so doing, we will see the hopeful beginnings of the evolution of the university, from a slow-paced, essentially reactive

J. T. Kenny (ed.). *Research Administration and Technology Transfer.*
New Directions for Higher Education, no. 63. San Francisco: Jossey-Bass, Fall 1988.

entity to an increasingly proactive force and catalyst in an emerging information society.

A Retrospective View

Two important concepts must be noted as we look at the university's historic response to external needs. These are *time lag* and *time compression*. By way of illustration, the printing press was invented in the late fifteenth century, but the Industrial Revolution did not occur until nearly three hundred years later. The electronic computer was introduced in 1945 and was quickly followed by the dawning of what we now call the information age. Here we see examples of lag and compression. In the latter case, the compression of time between related developments demonstrates the significant diminution of time available for response to changes in the external world.

A historic time lag is evident in the 1862 legislation designed to stimulate agricultural development. The Morrill Land-Grant Act was enacted over two hundred years after this country had established agriculture as its primary industry. Under this act, each state was given huge tracts of public lands as an endowment for agricultural colleges or agricultural departments in state-supported universities. Another time lag appears in the belated establishment of the first state system of free industrial education. For almost three centuries after the first settlement at Jamestown, Virginia (1607), this nation failed to promote the democratization of education by its unwillingness to pass laws providing free training for employment. It was not until 1905 that the Massachusetts legislature authorized Governor William L. Douglas to appoint a commission to investigate educational needs in the various industries of the state.

The start of the Industrial Revolution in the United States is traced to the 1830s and 1840s; the resulting university response occurred much later, in the 1870s, with the establishment of schools of mining and mechanical arts engineering. One might view the establishment of Johns Hopkins University and Hospital in 1867 as a proactive move in support of the development of a research-oriented institution. Nevertheless, the history of science and research indicates that the establishment of Johns Hopkins lagged far behind related developments in Germany and other European countries.

The initial trend toward transferring university technology began with the Hatch Act, in 1887. This act (also known as the Experimental Stations Act) appropriated $15,000 to each state to be used for establishing such stations. Its designated purpose was to aid in acquiring and disseminating useful and practical information in the field of agricultural science. The Agricultural Extension Act, adopted in 1914 (also called the Smith-Lever Act), provided a program of cooperative extension work in agriculture and home economics. It stipulated that such cooperative work should involve giving instruction and practical demonstration to people outside the universities.

With the advent of World War I, universities and the armed forces joined together in programs of military research. The wartime government created six war boards. These boards assumed the responsibility for adjusting American economic life to the necessities of the struggle. Along with the establishment of the boards, a most sweeping piece of legislation, the Overman Act, was adopted. These boards and the encompassing legislation were the vehicles that allowed the federal government to create laboratory-type structures involving the military and selected universities. It is interesting to note that although the laboratory-type relationship worked rather well after World War I ended, the structure used for the creation of these laboratories was abandoned. Physical laboratories were returned to universities, without the maintenance of formal agency or office to continue the effort.

The trend of modern-day intellectual migration began in the 1930s, when a number of academicians came from Europe to the United States. This migration was due largely to the pre–World War II conflicts that drove many Jewish and liberal-minded Europeans into American universities. These intellectuals brought with them a love of research and a fund of new knowledge.

A return to the trend of strong government association with universities came as a result of the United States's involvement in World War II. It took the invasion of Pearl Harbor, and the devastation of our naval fleet, for Congress to react once again. The government entered into a three-way association, including the military, universities, and industry. The laboratory-type structures associated with World War I were redeveloped. The most notable of these was the Manhattan Project. The Defense Department, with the University of Chicago and Oak Ridge National Laboratory, pooled its talents and energies to create the atomic bomb. This association laid the foundation for a national program of scientific/technological innovations, with the federal government playing a strong supportive role.

In the late 1940s, the Eightieth Congress initiated legislation that created the Atomic Energy Commission. This commission, designed to ensure domestic control and development of atomic energy, marks one of the first proactive moves made by the United States in preparing itself for a national research program. Under the commission, American scientists, primarily in universities, were contracted to further unlock the secrets of nuclear energy and to make their findings useful for peaceful as well as wartime purposes. The creation of the commission also gave birth to ultrasecret contractual associations between the federal government and universities and between the federal government and industry.

After the conflict, the scientific and research interest of the military, as well as of the Congress, began to phase out until 1957, when Sputnik changed for all time the importance of science, technology, and research. The inevitable combining of these topics into interrelated areas of knowledge provided another trend toward the reinvesting of resources in universities for future economic and social growth. The universities that took advantage of the government's renewed allocation of research resources in the late 1950s have become national and international leaders in the scientific fields. Certain innovations spawned by these advanced institutions of higher education were the electronic computer and the understanding of the electronic properties of semiconductors. Sputnik prompted a number of universities to reconsider their missions and give more emphasis to active research.

Since Sputnik, education in general, and higher education in particular, has exhibited strong reactive tendencies because of certain environmental pressures (the civil rights movement, urban education, foreign students, literacy, and so on). Moreover, from the late 1950s until the mid-1980s, the attention of the United States had been conditioned by a growing emphasis on military armament and related research. From an economic standpoint, this emphasis had become detrimental to the social programs of civilian America.

At approximately the same time that we were undergoing a change in attitude toward research, an educational revolution was occurring internationally. According to Rostow:

> The proportion of the population aged 20–24 enrolled in higher education in what the World Bank calls "lower middle income" countries rose from 3 percent to 10 percent between 1960 and 1982; for "upper middle income" countries the figure increased from 4 percent to 14 percent. For Brazil, fated to be a major actor in this drama, the proportion rose from 1 percent in 1965 to 12 percent in 1982. In India, with low per capita income but a vital educational system, the figure rose from 3 percent to 9 percent. To understand the meaning of these figures, it should be recalled that in 1960 the proportion for the United Kingdom was 9 percent, for Japan 10 percent [p. 841].

There has been, moreover, a radical shift toward science and engineering—for example, in India:

> The pool of scientists and engineers has increased from about 190,000 in 1960 to 2.4 million in 1984—a critical mass exceeded only by the United States and the Soviet Union. In

Mexico the annual average increase of graduates in natural science was about 3 percent, and in engineering 5 percent, in the period 1957 to 1973. From 1973 to 1981 the comparable figures were an astonishing 14 percent and 24 percent, respectively—an almost fivefold acceleration [p. 841].

The trend toward education and technology revolutions reemphasizes the fact that this country must be competent in the first endeavor (education) to be able to absorb and process in the second (technology).

Rostow also points out a development that will affect the role of the university in the twenty-first century. He states:

Still another somewhat related trend has been quietly at work reshaping the world arena of power over the past two generations: the relative decline in the economic power of the Soviet Union and the United States as Western and Southern Europe, Japan, and the developing countries of the Pacific Basin, and some of the more advanced developing countries in other regions have moved forward more rapidly. The combined gross national product of the United States and U.S.S.R. may have declined from about 44 percent to 33 percent of the global product between 1950 and 1980 [p. 839].

Science and technology are inextricably tied to economic power. As Rostow notes (p. 840): "The technologies that moved from invention to innovation in the mid-1970s include microelectronics, genetic engineering, a batch of new industrial materials, lasers, robots and various new means of communication." This association has four distinctive characteristics:

A close linkage to areas of basic science also undergoing revolutionary change; a capacity to galvanize the old basic industries as well as agriculture, forestry, animal husbandry and the whole range of services; an immediate relevance to developing countries to a degree depending on their stage of growth; and a degree of diversification such that no single country is likely to dominate them as, for example, Britain dominated the early stage of cotton textiles and the United States the early stage of the mass-produced automobile [pp. 840–841].

With Gorbachev's ascent to power, Russia has been moving away from an "armed camp" philosophy. Again, the United States in general, and American universities in particular, are left with the dilemma of reacting to global political pressures long after the fact. We are just beginning to understand that the future quality of American life will depend on the ability of the university to respond to the challenge of educational and technological change. The United States government and American higher education now have only a very short time to respond to these changes, while earlier we had relatively long periods of time to react. As these dual revolutionary trends progress, American universities are being forced away from the historically continuous reactive mode and toward the proactive mode of meeting society's needs. These revolutionary trends are dictating fundamental changes in the way in which universities function and signaling that there is a need for future structural changes.

Current Trends

The way in which we have historically done business has prohibited the United States from getting high-technology products to market with the same speed as the Germans and the Japanese do.

Figure 1 depicts the life cycle of an American product. Typically, research expenditures at the beginning of the cycle are greatest. University research laboratories are the most common sites of research, although industrial and federal sites also play a significant role. The industrial effort is greatest during testing and evaluation. The testing, evaluation, and commercialization of a product historically has been the total responsibility of industry. Expenditures of industrial time and money taper off toward the end of the product's life cycle, with maintenance and retirement efforts leveling out until the product is retired from the market.

This bimodal distribution of the effort expended during a product's life cycle clearly depicts one shortcoming in the way America approaches the commercialization of a product. That shortcoming is the gap between, on the one hand, university (or industry, or federal laboratory) research and, on the other, industrial testing of a viable product. This model can be applied to almost every mass-produced product in this country. The United States needs to develop a mechanism or process to make effort unimodal. Until this is done, we will continue to lag behind countries that have demonstrated their understanding of the continuum of effort all new products require.

The educational revolution taking place in today's world allows traditional technologies to be absorbed into other cultures. Other societies can now produce goods and services over which the United States historically had a monopoly. An example is the traditional smokestack industries leaving the United States and relocating in such countries as Taiwan and Korea. Competitiveness has become the driving force, which

Figure 1. Life Cycle of a Product

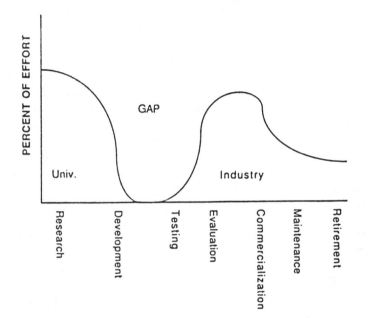

has caused us to develop institutional structures that will enable the gap to be reduced. An example of this competitive posture is the Omnibus Trade and Competitiveness Act of 1987. This legislation is designed to improve manufacturing and product technology and states that the federal government should maintain a laboratory system for national technology and outlines what form it should take.

This institutional structure goes a long way to fill the gap in the product cycle. What needs to be added, however, is a push from universities' research and development and a pull from marketing. The establishment of the technical laboratory, incorporated with a technological push and a marketing pull, is intended to change the Department of Commerce from a passive to an active organization. Practically all states are now addressing this need with the establishment of economic development departments at the cabinet level. These departments entice new industries to these states and help new businesses develop. A significant number of states also have international trade offices to assist in the exporting of goods and services.

As for universities' joining with state and federal initiatives, the academic response has been the establishment of multidiscipline centers. Universities are culturally, economically, and technologically oriented toward the departmental structure. To encourage multidiscipline centers, funding sources have instituted block grants that require close interaction among various university departments and colleges. This has resulted in the proliferation of multidiscipline centers designed and controlled by matrix-management techniques. The reason these centers now appear is because the university so far has not been able to change its rigid departmental and curricular organization. The University of Colorado has created twenty centers in the recent past. This current trend, it is estimated, will lead to 60 percent of federal funding for research and development being channeled through matrix-management centers.

The Department of Defense is a catalyst for developing centers. It should be noted that 75 percent of federal research and development money is related to meeting defense needs. This percentage has been steadily increasing throughout the 1980s, despite the fact that the United States has become less and less efficient in the civilian marketplace. The university has attempted to join with industry in the establishment of industrial parks. These parks have been primarily placed near incubator systems for new companies and established high-tech companies. Unfortunately, many of these industrial parks have been nothing more than real-estate projects, with relatively little association with nearby universities.

The current trend is to recognize that technology is just as important to economic success as capital, labor, and materials. The idea that high tech feeds commerce has changed the unit of power from military strength, with the United States and Russia dominating the rest of the world, to economic strength. For example, Japan has strong economic power, but little if any military power. This change in the unit of world power has caused a diffusion of strength away from the United States and Russia and into other economically developed countries. This fact is causing Russia to undergo fundamental changes in its governmental institutions.

These fundamental changes are being accelerated by the growth factor (10^6 per decade) in the tools of the information age. This factor has been derived from measuring the increased capacity to store, retrieve, and assimilate information from state-of-the-art computers and their networks. The response to this influence in most universities has been the creation of a management structure that begins at the vice-presidential level to deal with the strategic, tactical, and operational requirements of these technological advances in computer science. This has required universities to retool their faculties to meet the academic needs of their incoming students. Most institutions have some form of computer-learning process targeted for faculty. Current interest should be focused on the use of artificial intelligence and other fast-moving topics in computer science.

The competitive factor will drive more and more universities to refocus their emphasis on research and technology transfer. Because of increased understanding among the general public of the strong relationship between research and new commercial products, people tend to support research in the universities. This is particularly significant in the case of public land-grant universities. The magnitude of this polity influence is going to be as great as the influence on defense when the Russians first launched Sputnik. The new aspect of competitiveness is the emphasis on the transfer of technology from academic environments into the private sector. Many pieces of federal legislation enacted in the 1980s have focused on technology transfer. These legislative responses have led to the creation and expansion of structures within universities to deal with intellectal properties. Another area of interest is the model of agricultural extension, which expands the roles of county agents and community experts beyond agricultural areas. Many state universities now employ extension agents and experts in engineering. Unfortunately, time lag occurs here, too, and the response is all too often reactive.

Our discussion of past and current trends in American higher education might suggest that, at least for this sector of American society, the "soft landing for the Cold War," as envisioned by Rostow (1987), may be as easy as landing softly on a bucking horse in a competitive rodeo.

The Compressed Future

The first and foremost projection we might offer is that tomorrow's campus, and all of society's institutions, will have to function in a very dynamic environment. In the past, we have been accustomed to long periods of minimal change in our institutional structures. The forces of change build to a breaking or disruptive point in the original structure, causing a short period of instability, which ushers in a new and slightly altered level of the status quo. Certain structures have historically been normatively invariant over time. For example, religious doctrine and institutions have tended to be quite static. With rapid advances in biological knowledge, elements of religious norms are increasingly challenged by new ideas and values, but this pressure does not always prompt doctrinal change.

A prime example of an institution that evinces flexibility and whose doctrine has varied over time is the United States government under its normative and legal framework, the constitution. Interpretation of the constitution is very time-dependent (for example, the 1954 *Brown v. Board of Education* decision, affecting civil rights and education). Many of our institutions will have to adapt to a changing environment while retaining their historic identities. The distinguishing characteristics of an institution today may not be the same characteristics that will describe that institution in the future. To illustrate, an eligible voter in the 1790s was a white male landowner. Today, an eligible voter is a citizen who is eighteen or older. This change in the definition of a voter evolved over two hundred years. In the twenty-first century, such institutional changes may occur in less than a single decade. Thus, there will be a need for time-dependent descriptions of institutions.

An increase in the time-dependence phenomenon will be seen in the changing management structures of our institutions. Research management in universities has been somewhat rigid, and to circumvent this problem, universities have instituted matrix management. This process is really a stopgap measure but in itself will initiate needed changes in management structure. One unique idea is derived from a Kellogg Foundation grant recently made to the University of Michigan. Under the terms of the award, 1 percent of the institution's total operating budget is reallocated, or "reprogrammed," into new initiatives for solving problems that the current structure cannot effectively handle. This process encourages the institution's ability to affect the environment. With this reprogrammed money, the institution is obliged to ensure that the new initiatives are better at using the new knowledge that is being generated. This requires an understanding of the dynamics of the external environment and a commitment to strategic planning. This level of planning must become an integral part of the institution's management structure, so that efforts directed at innovative change become the rule, rather than the exception.

A major economic pressure confronting the United States, as pointed out by Rostow (1987), is that more and more developing countries will soon be able to produce and market goods and services internationally. This pressure will cause significant changes in the way in which commerce is conducted in this country and abroad. The United States will be forced to export more, and countries like Japan will have to consume more. United States institutions will hve to make it as easy to trade overseas as it is to trade across Main Street. According to Mikuni (1987, p. 31), "Japan is undergoing rapid change. Differences will remain, especially cultural contrast with the United States, but the gap between our *modus operandi* will close to some degree with the developing of a free market economy. These changes are perpetuated by the fact that historically, in Japan, industry and exports have been the great gods. Today's private sector would prefer that the consumer be king as in America."

In today's economy, the United States sees approximately 10 percent of its gross national product as exports. By the turn of the century, the United States will need to increase its exports fourfold. This will be reflected in how the federal government allocates money. The Department of Commerce currently spends $2 billion per year, and the Defense Department spends over $300 billion per year. The shift in the unit of power away from militarism and toward the trading side of the economic continuum indicates the need for the Commerce Department to be spending more equally with the Defense Department. By the year 2000, the Commerce Department should be spending on the order of $30 to $40 billion, and the Defense Department should be spending on the order of $80 to $90 billion (in 1987 dollars). The precipitate dollars might be channeled into education, labor, energy, health, and welfare, as well as into balancing the budget. This reemphasis on economic growth and restructuring of institutions is not only taking place in the free world but is also the centerpiece of Gorbachev's policy direction for Russia and its satellites. China is also moving toward a freer economy. This confluence of new national goals, country to country, may herald unprecedented growth in world trade for highly valued products and services.

The Soviet Union and China tend toward a collective response to external demands in the environment. This has made them less commercially competitive than the West. In much the same way, independent responses on the part of U.S. industry (with minimal government support) have caused us to be less competitive. In this independent mode, we tend to focus on profit, whereas we should exhibit a better balance between near-term and long-term objectives. Although our cultures and our historic governmental structures have been different, environmental forces are now causing all of us to assess whether the collective mode or the independent mode will be most beneficial. There will be times when this determination is not important, and we will return to some of our cultural and institutional differences; but, for example, if it becomes clear that a collective mode (government plus business) is the best response to a given problem, then the institutions that must make the response must be flexible. Obviously, the way in which we move from basic research to commercialization of a product must change in order for us to be competitive with Japan and other industrialized countries. The future trend in the United States must be toward a more collective response on the part of universities, government, and industry.

To remain abreast of various developing trends and influence their direction, we will need to foster holistic strategic planning for our diverse institutions. The federal government, state governments, universities, and industry must identify and pool their resources to meet common goals. This means that a state that wishes to build a particular type of high-tech industry will have to assist in building an appropriately educated labor force and provide for the development of the technological base. This will require universities to be involved at the same level of planning as state government and related industries are. This type of strategic planning is found today in Japan, where government and industry have joined to acquire the appropriate technology base. To date, Japan has not fully included its universities in strategic planning, but Japan is closer than anyone else to addressing the demands that future trends will dictate. The United States will need to have a well-articulated and well-developed industrial and economic policy, planned and discussed by universities, state and federal government, and appropriate industries. In today's environment, we have allowed our mature industries to stagnate in their own quest for a larger share of short-term rewards. The Russians, in contrast, are going to have to give more autonomy to local institutions and industries. All the Soviets' planning is central. This has resulted in the reduction or elimination of individual contributions. The United States needs to review these two extremes and develop a better mix, so that the resulting changes in the environment can be relatively painless.

During the Industrial Revolution, society developed one technological invention after another, without considering their social impacts. That world envisaged the move from an agrarian to an urban society, and then to a suburban society, leaving behind poverty, illiteracy, and unemployment. Change without pain means acquiring the ability to solve sociotechnical problems, without solving just the social or just the technical components in isolation. We are moving toward a more sophisticated society, driven by a generation of more sophisticated technologies. For all citizens to enjoy this country's prosperity and have equal access to it, the gap between the haves and the have-nots must be significantly reduced. The future will force society to move toward appropriate combinations of collective and individual initiatives.

A new trend in leadership will be needed. Decision makers will become the critical variable. One important characteristic of a future leader will be his or her ability to make the transition through success. Usually, we experience success and continue to apply the same methods that produced it. For our institutions to experience accelerated growth, leaders will need to develop new problem-solving capabilities to achieve new successes in a changing environment. What was successful at one point may prove disastrous later. Leaders will have to read and influence the environment, to recognize changes that have taken place and are to take place.

Ability to understand the changing environment means lifelong learning. In the past, one could learn problem-solving methods that would be useful and valid for a lifetime. Methods were organized around single sets of rules, which were applicable only to problems in a static environment. In today's universities, we focus on educating people eighteen to twenty-two with a prescribed body of knowledge, an algorithm for continued learning, and methods for solving problems. Within a few years, students' current knowledge will be outdated. To understand the inadequacy of the algorithm for continued learning, consider what would happen if turn-of-the-century Newtonian mechanics were employed to solve problems in atomic physics, without the intervention of quantum mechanics. Extending what is known about Newtonian mechanics would not enable one to understand atomic physics. By the same token, when we try to solve sociotechnical problems only with technical tools and reach only technical solutions, we are probably selecting the wrong answers. Lifelong learning will be necessary in the future to solve the many types of problems we will face.

As we go through this learning process, information will need to be transferred from one institution to another with increasing speed. A new trend in the sharing of intellectual human resources will appear. A professional's upward mobility not only will be assessed at his or her own institution but will also depend on his or her success in government, industry, and academia. Thus, it will be very common for an individual to have three distinct lines of mobility prospects, instead of the traditional one. To facilitate this sharing, institutions will need to have missions that not only are internally self-serving but also are expanded to include connections with the environment.

The University of the Next Century

Strategic planning in most major universities today takes place every three to eight years. Planners tend to investigate the environment and react on the basis of limited changes in institutional structure and limited vision. The most significant changes that take place in the university tend to be unplanned changes, with little or no forethought. The driving force is sometimes some significant change in the external environment, which leaves the university to solve problems as best it can. Examples of environmental changes have been the early Soviet space initiatives, the Nixon administration's cutting 40 percent of the NASA budget in one year, the drought of 1980–1981 in the Midwest farm belt, the economic repercussions of the lagging American automobile industry, and the accelerated cost of gasoline with the subsequent decline in oil prices. We have been inefficient in the past at identifying major trends and their probable impacts on academe. The livelihood and development of the future university will rest on its ability to read, influence, and be proactive in a rapidly changing environment. Thus, universities will need to establish futures or think-tank groups. These groups could be responsible for forecasting trends, predicting the impact of trends on the university, and designing means by which threats to the university will be turned into opportunities. Traditionally, only a few individuals have considered and projected the direction of the university (the president or a governing board). Today there is very little assistance available to define the path the university is traveling; the futures group will illuminate the path. In order for the group to be effective, it will have to develop methods and tools with analytical and predictive utility. Proper use of the futures group will enable the university to become considerably more proactive.

Another structure that will need to appear in research universities because of the technology revolution will be the *environmental window*. There will be so much new information developed outside particular institutions that internal entities will have to be responsible for collecting, packaging, and comprehensibly transmitting it to faculty, students, and administrators. Operational units that will be designed to collect, package, and train are what we are labeling the environmental window. We have seen the introduction of technologies that may be created outside the university but are not absorbed by it (for example, computers and, more recently, artificial intelligence). These are technologies that penetrate many disciplines within the university but are late in being incorporated as research topics. In the case of computers, most universities have by now established centralized computing centers, which not only house computers but also train faculty in how to use the new computer technologies. We forecast that these computer centers will be combined with libraries and expanded into information centers. One unit of the information center will be the environmental window. Some universities have learning centers devoted to improving teaching capabilities. These learning centers, in their own right, will be absorbed or expanded into such windows.

One means by which faculty will be kept close to the frontiers of knowledge on a wide variety of topics will be sabbatical leaves. The faculty member of the future will need to spend a significant amount of time in a cross-section of universities, industries, and governments (local, state, and federal). The ways in which promotion and tenure now occur will have to be changed to accommodate the faculty member's experiencing this "revolving door" trend. This trend will also emphasize the need for coordinated strategic planning among government, industry, and academia. This proposal will have a very important impact on how long-range planning survives changing governmental administrations. University faculty, and institutions themselves, will have to maintain vision, direction, and leadership by keeping governmental and industrial partners well informed. For this scenario to take place, faculty must first be informed, and then precise systems must be established to maintain consistent levels of information dissemination and sharing.

The preceding analysis has supposed that faculty members and students of the future will be capable of solving sociotechnical problems. Enhanced creative thinking, and the understanding of complex issues associated with rapid "change without pain," will be important components of the research university of the twenty-first century. Managers and leaders of the university will have to be at least as enlightened as faculty about solving sociotechnical problems. This will require more formal research and more formal training of those who want to be managers and leaders. Universities lag far behind industry and government in training. We lack knowledge about the development of leadership qualities, the characteristics that make successful leaders, and a sane process by which potentially successful leaders can be chosen. At present, our processes of choosing leaders are no more correlated to leaders' success than random selection would be. The measures that we currently employ to select university presidents, for example, are not consistent with the tasks and obligations that will confront the university of the future. Successful leadership will mean more than stabilizing and perpetuating the institutional status quo; it will mean keeping up with changes in the environment and taking proactive initiatives to influence outcomes positively.

Some of the changes that will occur in academe in the twenty-first century will require dynamic leaders and enlightened faculty. An example will be the quantum increase in the institution's involvement in international interactions. A greater number of foreign students will attend American universities, and there will be increases in the number and tenure of American students who attend foreign universities. This level of exchange will promote more focused attention to cultural studies, as well as to the forces influencing expanded international trade. Increased exchange cannot occur unless institutions have internationally attuned faculty.

Changes within the university and in the external environment will require that the structures of semi-independent departments be changed so as to reflect more collective modes of management. These collective modes, in turn, will both reflect and influence the external environment. Since the milieu will be evolutionary or revolutionary, management structures must change to accommodate change dynamics. One such current change in public education, which will accelerate in the future, is the movement from state-supported to state-assisted universities. This change is being driven by federally and industrially sponsored research and development. The so-called public research university receives approximately two-thirds of its operating budget from the state government, with the remaining third coming from a combination of federal government, private philanthropy, tuition, and private industry. By the twenty-first century, the funding mix will be the reverse of the current proportions: one-third will come from the state, and two-thirds will come from other external sources. The obligation of the state will not diminish, but the university will grow and assume new tasks and responsibilities associated with the environment. One example of such a task is the university's obligation to provide lifelong learning for faculty, students, administrators, and a broader public. The current mode of training university students takes approximately five years. Taking into account that a person in the future will live to be almost one hundred years old and be gainfully employed for fifty or sixty years, we are looking at a tenfold change.

It is important to remember that the twenty-first century is just a little more than a decade away. The starting points for instituting the necessary changes will be the university's research officers and its managers (insofar as the latter group interacts regularly with government and business and is already heavily involved in the assembly and transfer of new knowledge).

As we approach the "soft landing of the Cold War," which Rostow (1987) tells us is being caused by the revolutions in education and technology, the university will be placed in the role of change agent. This new role will open new dimensions and impose new obligations. Universities of the future will be taking uncharacteristic risks in charting new paths and improving innovative productivity. Times are changing, however, and an evolution is taking place that moves us beyond even this focus. Today the rapid expansion of knowledge, and its absorption into the changing world culture, is our newest frontier.

References

Mikuni, A. "Market Forces Urge Japan Toward a Consumer Society." *The Wall Street Journal*, Nov. 16, 1987, p. 31.

Rostow, W. W. "On Ending the Cold War." *Foreign Affairs*, 1987, *64* (4), 831–851.

Thomas C. Collins is vice-provost for research at the University of Tennessee, Knoxville.

Sheadrick A. Tillman IV is associate vice-provost for research at the University of Tennessee, Knoxville.

Changes in federal patent policy, and the coincident loss
of competitiveness and worldwide market shares by American
industry, have spurred increased interest in university-
developed technology and technology transfer programs.

The New Frontier
of Technology Transfer

Gerald A. Erickson, Donald R. Baldwin

There is an exciting new direction in American higher education. It is commonly referred to as *technology transfer* and is increasingly the subject of attention from academic institutions, industrial corporations, state governments, and the federal government. All indications are that the topic will be an important one for several years to come. Speedy development of the concept and the refinement of effective models to achieve it are important to the United States' competitive position in the world marketplace.

The notion of transferring knowledge is certainly not new to academic institutions. The traditional methods of disseminating the results of university research have included the education of students, the publication of books and articles, and the presentation of papers at academic and scientific meetings and conferences. Those methods continue, and they are largely unaffected by the new one, technology transfer, which is at the heart of this new frontier. For the purposes of this chapter, *transfer* refers to research results converted to commercial products, processes, or services covered by patents or licensing arrangements. Before discussing this phenomenon, it may be useful if we note some of the primary reasons why American universities have not historically participated the more commercial forms of this practice.

J. T. Kenny (ed.). *Research Administration and Technology Transfer.*
New Directions for Higher Education, no. 63. San Francisco: Jossey-Bass, Fall 1988.

The Soviets' launching of Sputnik spawned a vigorous American commitment to science and technology. The federal government itself took a leading role and during the 1960s and 1970s allocated ever-increasing sums of money to support basic research by the country's colleges and universities. Research grant awards were based on a scientific peer-review process. There was plenty of money, the application procedures were easy, and, at least until the early 1970s, the postaward administrative requirements were not onerous for faculty investigators and grantee institutions. Given these circumstances, university-industry research collaborations diminished, in part because industry funds were not needed to support university research, and in part because industry's rights to results were clouded by actual or potential commingling with federal research funds. In this growing separation between the academic and industrial research communities, a critical bridge was lost for the continuous and speedy transfer of research results from the academic laboratory to the marketplace.

This problem, and its growing seriousness, began to come into focus during the 1970s and eventually led to a series of steps designed to revive and strengthen university-industry interactions. Basic technologies produced by universities in the 1970s lay largely undeveloped by American companies and often were commercialized by foreign firms, especially Japanese companies. This led quickly to an erosion of American technological preeminence in the world marketplace. To correct this, government, industry, and academia began to form new partnerships designed to focus on research with apparent commercial promise and to minimize barriers to the expedient transfer of results. One early response by the federal government, through the National Science Foundation, was to sponsor five innovation centers at campuses across the nation. The role and direction of these centers, along with major industrial and economic factors related to the concern over American competitiveness in international and domestic markets, were reviewed in a 1978 symposium (see Hammond, 1978; Ramo, 1978). While it was clear that the innovation centers were not the full solution to the problem of technology transfer, their efforts, the ideas they sparked, and the concepts they tested confirmed the wisdom of developing stronger university-industry relationships.

The trade situation has continued to erode during the 1980s, with U.S. companies losing market share both within and outside the country. As a result, the annual balance of payments went from a small surplus of $5 billion in 1980 to a hemorrhaging loss of over $150 billion per year from 1985 through 1987. To regain market share and eliminate this deficit, American industry must improve its ability to compete at home and abroad. An important factor will be the incorporation of new technological advances from the nation's universities. The key threads in this overall fabric of change are discussed in this chapter, beginning with a look at the pertinent federal patent and invention policies.

Important New Federal Factors

Starting about 1980, federal encouragement of technology transfer by universities increased, because of important changes in two areas: federal patent policy revisions, and the federal courts' support of damage awards for patent infringement. Before 1980, patent policy inhibited commercial development of federally sponsored university research results. Policy changes that allow universities to retain ownership of patents based on federally funded research have increased the number of technologies available for commercialization. The ability to profit from these patents encourages universities to be more aggressive in licensing.

Revised Federal Patent Policy. Before 1980, the federal government did not have a uniform, governmentwide patent policy. Individual federal agencies were allowed to formulate and implement their own policies. These circumstances did not encourage university faculty researchers or institutional administrators to commit either attention or resources to vigorous technology-transfer programs. Further, prospective industrial licensees were generally skeptical of technologies based on federally funded research. They feared that titles or licenses granted by universities would be clouded by government rights, and that their ability to exploit the technology fully and realize its potential value would be diminished.

This situation changed dramatically when Congress enacted Public Law 96-517, the Patent and Trademark Amendments of 1980, giving universities the right to retain title to federally funded inventions they had developed. The original act was improved when Congress amended it in 1984, through Public Law 98-620, to extend its coverage and remove or ease some of its restrictions. A recent study by the U.S. General Accounting Office (GAO) (1987) indicates that the title-rights provisions of these laws have had a significant positive impact on universities and small businesses. University administrators surveyed by the GAO indicated that Public Law 96-517 had been important in stimulating business sponsorship of university research, which grew 74 percent, from $277 million in 1980 to $482 million in 1985. The same administrators also noted that the Public Law 98-620 amendments, which removed licensing restrictions, will be particularly significant, because the new provisions will make businesses even more willing to license university-developed technologies. It is clear from the GAO report (1987) and other recent studies that the passage of Public Law 96-517 motivated many universities across the country to implement new or more vigorous technology-transfer program.

Nevertheless, there were other factors. By 1980, many states had taken a keen interest in economic development. They generally recognized the important role that universities play in economic development and the relevance of strong university-industry link to fostering technology transfer and the creation of new jobs. It is also somewhat pertinent that in the early 1980s many public and private universities were having severe financial problems and were very attentive to any prospect of new funding sources. The realization that technology-transfer programs might be one such source required little imagination. Indeed, the national news media were carrying a growing volume of stories about university faculty members' and some universities' having received handsome financial rewards for new biotechnology inventions developed in university laboratories.

Response of One University. Experience at the University of Washington serves as a concrete example, and its response to this fabric of circumstances seems to be typical of that of many universities, both public and private, across the country. In the fall of 1981, the university provost appointed a task force of senior faculty members and administrators to consider the full set of issues related to technology transfer. The group was asked to recommend whatever policies, organizational changes, and procedures were deemed appropriate to ensure an effective program consistent with the new federal patent policy, other related national objectives, and the university's own mission. The task force gathered and analyzed patent and copyright policies from selected universities, took testimony from faculty throughout the campus, and met with local industrial leaders. The effort took approximately nine months and eventually led to four major recommendations.

1. Revise the patent policy to reflect an explicitly positive attitude toward transfer of research results to the private sector, and include incentives to encourage that transfer.

2. Adopt a copyright policy as a companion to the patent policy, so that the two policies together cover all tangible and intellectual properties.

3. Establish an office of technology transfer to administer the patent and copyright policies, coordinate university-industry interactions, and give greater visibility to technology-transfer functions.

4. Adopt a policy concerning employees' involvement with commercial enterprises, to clearly encourage appropriate involvements and to provide uniform guidelines.

After consultation with the faculty senate, the provost and the president accepted all four recommendations and moved to implement them. Early in 1983, the Office of Technology Transfer (OTT) was established, and a revised patent policy and a new copyright policy were adopted. The policy on employees' involvement with commercial enterprises was taken up later and adopted in 1985. The early results of the university's new program are shown in Table 1, which demonstrates that dramatic results are possible. The scope and timing of such results will vary from one institution to another, depending on the volume and type of research, as well as on the expertise and resources committed to new or expanded technology-transfer programs.

Table 1. Selected Activity Indicators

	Fiscal Years Ending June 30			
	1984	1985	1986	1987
Disclosures				
Invention disclosures	33	47	62	70
Pool of identified inventions	90	137	167	228
License Agreements				
New licenses signed	1	11	16	20
Total licenses in effect	28	37	50	67
Total options to license in effect	7	4	7	8
Licenses producing income	12	11	21	32
Income				
Royalty and license fee income	$ 35k	$ 76k	$360k	$397k
Sale of biological materials[a]	$ 6k	$ 20k	$ 17k	$141k
	$ 41k	$ 96k	$377k	$538k
University-Industry Interactions: Research Agreements with Industrial Sponsors				
number	312	379	419	424
value	$7m	$12m	$11m	$15m

[a]Approximate figures; not reported separately until FY87.
Source: Graduate School Office of Technology Transfer, University of Washington, 1987.

Federal Court Decisions. A more liberal federal patent policy has little meaning if the patents that result are not enforced. The value of patents, in their ability to exclude competition, has increased with recent decisions by the federal courts in infringement cases. The shift came with the establishment of the Federal Circuit Court of Appeals in 1983 (Whipple, 1987). The tone of the court's decisions was more propatent, and it imposed more damage awards, than had been the case in previous courts of appeal. A coincident Supreme Court decision in 1983, in *General Motors* v. *Devex,* made prejudgment interest awards the rule, rather than the exception. Decisions in the new court of appeals during the following year set a tone for deciding in favor of the patentee on uncertainties in appropriate damages, and for increasing punitive damages beyond amounts that would have been reasonable royalties.

Important Factors for University Research and Development

Loss of international market share by American companies, and the increased competitiveness of foreign firms in U.S. markets, have forced U.S. firms to seek ways to improve. Many of their hopes appear pinned on using new technology to compete more effectively.

Reduced U.S. Dominance of International Markets. Because of the lack of competition in international markets while Japanese and European industries were being rebuilt after World War II, American industry was not pressured to stay lean, mean, and aggressive during the 1950s and 1960s. The result of the earlier lack of competition and reduced American competitiveness was a dramatic drop in market share in major industries, worldwide, beginning in the early 1970s.

Competitiveness and New Technology. Jonas (1987) highlights the basic concerns that have developed over the last fifteen years, as well as the importance of technology for improving U.S. productivity and competitiveness. The key concerns include maintaining a strong manufacturing capability in America, using new technology to improve manufacturing and productivity, capital investment with a long-term view and commitment, an educated work force, and good management.

Reemphasis on the Importance of Industrial Relations. University-industry interaction will continue to increase because of mutual benefits. Effective transfer of universities' research results to industry helps industry become more competitive, and research support from industry offsets decreases in federal funding for academic research. For example, industrial support for research at the University of Washington has grown from $7 million in fiscal 1984 to $15 million in fiscal 1987. While this is a significant amount, industrially funded research still constitutes only about 6 percent of the university's total research funding ($253 million in 1987). This level of industrial funding is equal to the average among the top twenty-five American research universities; Massachusetts Institute of Technology (MIT) and a few other private universities do much better. Ken Smith, MIT's vice-president for research, reported that industrial funding at MIT had grown 20 percent per year since 1976, a rate faster than any other source. Industrial funding was $36 million in fiscal 1986, about 15 percent of all research monies ("Corporations on Campus," 1987).

Small, single-sponsor research contracts or multisponsor research consortia appear to be common, while large, multimillion-dollar, single-sponsor programs seem rare. The Monsanto–Washington University (St. Louis) program is a notable example of the latter. The University of Washington experience includes a couple of multiple-sponsor examples. The Center for Process Analytical Chemistry (CPAC) consortium was created in 1984 at the University of Washington with National Science Foundation (NSF) backing and over twenty industrial sponsors. The CPAC 1987 funding rate is about $1 million per year. The Washington Technology Center (WTC), an entity designed to coordinate state and other research support in selected fields, was created by the Washington State legislature in 1983. The WTC 1987 funding rate is about $3 million of state support, plus $8 million of federal and private industrial support, and the rate is growing.

Federal Programs and Tax Incentives. In addition to the new federal patent policy already mentioned, the federal government has taken a number of other steps since 1980 that also foster university-industry relations and encourage technology transfer. Antitrust laws have been relaxed to allow firms to form collaborative research ventures, often with academic scientists, to solve technological problems common to their industries. The Justice Department now sanctions such cooperative research, with the understanding that cooperation will end and competition resume as soon as research is turned into product. Forty such research consortia have already been formed, and more are being planned. The University of Washington's CPAC is another example. The feasibility of CPAC was confirmed with assistance of a planning grant from the NSF, and the center was implemented in 1984 with the backing of a five-year grant. By its third anniversary in 1987, CPAC had thirty industrial sponsors, an impressive list of scientific publications, and a growing volume of invention disclosures.

Tax laws have been amended to provide incentives for research and development. For example, 1981 legislation provided corporations with a five-year tax credit for research and development. The 1986 Tax Reform Act extends a modified version of the tax credit for three additional years, through December 1990. The credit provisions are structured to encourage corporations to donate research equipment to universities and to provide funding for university research. These tax incentives have stimulated many university-industry interactions that probably would not have occurred without them.

Changes in the U.S. Patent and Trademark Office are also relevant. Recent automation has streamlined the operation and reduced the average wait for patent issuance. Further, the number of patent depository libraries was expanded from seven in 1985 to sixty nationally.

Economic Development Strategies. Almost coincident with the shift in federal patent policy, states have increased support for small-business growth and higher education, with the realization that both are vital to a healthy, growing economy. Alabama, Florida, Texas, Arkansas, North Carolina, and South Carolina are pouring unprecedented sums into public education in general and into vocational-technical education in particular ("Growing Places," 1986). Ohio has established the Thomas Edison Program to match academic resources and researchers with support systems for launching commercial technology ventures. Indiana has established a $135 million fund to invest state-sponsored seed capital for high-technology start-ups in thirteen disciplines. Iowa has recently allocated $18 million for a biotechnology center that combines the resources of both its major universities. Pennsylvania has the Ben Franklin Partnership and the Pennsylvania Technical Assistance Program, which support technology use and economic growth in the state. Michigan invests state pension funds in venture capital and uses a strategic fund that contributes to bank loans for businesses that would not qualify for conventional loans. Michigan's and Pennsylvania's programs were created with the aim of supporting local business development.

Economic Development Factors and Attitudes. Birch (1981) indicates that firms with fewer than twenty employees accounted for 66 percent of all new jobs between 1969 and 1976, companies less than five years old created 80 percent of all U.S. jobs during the same period, and the Fortune 500 companies created virtually no new jobs in the 1970s. Birch's work has had substantial influence on thinking about economic development at the state and local levels in the state of Washington.

The concomitant shift to support of local businesses has an influence on the focus of universities' technology-transfer efforts. In addition to the licensing of existing businesses, serious consideration is given to start-up opportunities for businesses in Washington by the Washington Research Foundation, an agent for the University of Washington and other state research institutions. WTC has implemented a "Washington First" policy, which provides access to newly available technology from WTC to Washington-based businesses, before national or international marketing efforts are undertaken.

Importance of Higher Education Centers. Birch (1981) highlights the importance of new businesses to the national and regional economies and identifies the dominant role that these businesses play in job formation. The importance of higher education centers has also been recognized for the formation of new businesses that can compete on the basis of technological advantages. Higher education centers also provide a base of skilled and knowledgeable individuals.

The top areas for entrepreneurial activity and new venture capital have been the San Francisco Bay Area and Route 128 around Boston. These areas are also noted for their many excellent universities. Strong economic growth is occurring in the Research Triangle Park area of North Carolina, which benefits from the contributions of Duke University, the University of North Carolina, and North Carolina State University. In the Puget Sound region around Seattle, strong growth in the biotechnology community has occurred, fueled largely by world-class researchers and the development of new technologies at the University of Washington and the Fred Hutchinson Cancer Research Center. In the Salt Lake City area, new business growth is being spurred by a more aggressive technology-investment attitude at the University of Utah.

A recent article addressed to job seekers in the 1990s says, "Go where there are large concentrations of brains, plus cultural amenities that attract bright and creative people. . . . The most promising spot of all seems likely to be near prestigious, research-oriented universities"

("The Economy of the 1990s . . . ," 1987). Recognition of the importance of higher education to job formation and economic growth has begun affecting states' strategies.

Trends in the Academic Community

The last five years have seen substantial growth in the number and scope of universities' technology-transfer programs. This had led to increased activity in professional organizations that serve the needs of those who work in university and industrial technology-transfer offices and in the level of activity in established university programs.

Professional Associations and Societies. The building interest in technology transfer is reflected in the increasing membership of the Society of University Patent Administrators (SUPA) and the Licensing Executives Society. Both organizations are attempting to meet members' needs through expanded services, such as a directory of journals and members that was recently implemented by SUPA, and more regional, national, and international meetings, conferences, and workshops.

Other organizations have also expanded their previous charters to embrace technology transfer. For instance, all regional and national meetings of the National Council of University Research Administrators now include major sessions and workshops on various aspects of technology transfer. The National Association of College and University Attorneys (NACUA) has recognized the need to acquaint its members with the legal elements of the topic. In 1985, NACUA's annual meeting included only one session on technology transfer, but in 1986 it included four sessions. Various private organizations have also recognized the market for courses and workshops on technology transfer and are vigorously pursuing that market.

Operating Models. While there are many variations, the two basic approaches to handling patents, licensing, and industrial liaison activities in academic institutions are in-house offices and outside-agent models. There are pros and cons for each approach, and the right choice is generally related to the particular circumstances of a given university. Assuming that it is adequately staffed with competent and experienced people, the in-house office will be more responsive to the institution's needs. The price of that responsiveness may be very high to acquire patent attorneys, market analysts, industrial liaison experts, and the secretarial and other staff to support them. In a new program, investment in such staff is a relatively high risk, since there is no guarantee that the investment will eventually be recovered by successful patenting and licensing activity. Even if it is recovered, the recovery cycle is likely to be five years or more, because it generally takes that long to move from invention disclosure to significant sales of licensed products.

Given the large up-front investment and the relatively high risk of that investment, many universities opt for an outside-agent model. It ensures immediate availability of the required legal, marketing, and business expertise. It avoids the need for substantial up-front money to cover evaluation, patenting, and marketing costs, pending much later recovery of those costs through fees and royalties. It also avoids the very real risk that the up-front expenses will never be fully recovered. What the university gives up in this model is some degree of responsiveness to its particular needs and some share of the income that must be allocated to the agent.

Some universities have tried to have the best of both worlds by using a combination of the two basic approaches. Again, the University of Washington may be typical of those in this group. The University's Office of Technology Transfer is responsible for administering the university's patent, invention, and copyright policies and for coordinating its research interactions with industry. The university has reserved to itself the right to undertake any and all patenting and licensing functions it chooses. The office is very modestly staffed, and it relies heavily on a few agents for patenting and licensing. This approach is designed to ensure responsiveness to the university's technology-transfer program while limiting the university's investment. In general, it may be prudent for some universities to start with the outside-agent model and gradually develop an in-house office as the program matures, experience grows, and income increases and becomes more predictable.

Stanford University and MIT are frequently cited as successful users of in-house offices. Their annual incomes from fees and royalties are in the range of $4 million to $6 million. That level of success is beyond the reach of many other universities. Stanford and MIT have very large, highly prestigious research programs. They also have extensive longstanding links with industrial sponsors and collaborators. Finally, they have made heavy investments over many years to develop their technology-transfer programs. Any university attempting to develop or evaluate its own technology-transfer program must be realistic about setting goals and, most particularly, about the time during which those goals will be achieved.

Licensing Agents with National Scope. The University of Washington has established relationships with two organizations that are nationally prominent in technology transfer, Research Corporation and Battelle Development Corporation. The advantage of using such organizations is the depth of experience and resources that they can provide to effect implementation and adoption of technology by businesses. The primary disadvantage is diluted attention, due to the number of institutions that such organizations serve or to internally developed technologies that compete for attention and resources.

Research Corporation. Research Corporation (RC) was founded in 1912 by Frederick G. Cottrell, inventor of electrostatic precipitator sytems for controlling industrial air pollution, and by Charles Doolittle Walcott, Smithsonian Institution secretary. The objectives of RC are to increase the availability of inventions and patent rights and to promote and enable scientific research and experimentation. RC provides grants for basic research in the natural and physical sciences from its endowment income and donations from other foundations and industrial companies.

Battelle Development Corporation. This organization is a wholly owned subsidiary of Battelle Memorial Institute. Battelle was incorporated in 1935 to encourage and develop discoveries and inventions and to license or sell technology to industry. Battelle receives invention disclosures from universities, industrial firms, technology-transfer organizations, independent inventors, and Battelle Institute's own laboratories, which conduct over $400 million in sponsored research annually.

Focused Nonprofit External Agents. In addition to working with licensing agents that have national, multi-institutional scope, several universities are working with agents that have more limited scope, focused on the university alone or on a closely associated geographical region. The Wisconsin Alumni Research Foundation (WARF) and the Washington Research Foundation (WRF) are two examples.

WARF. The Wisconsin Alumni Research Foundation was created in 1925 to develop the inventions of Wisconsin scientists and scholars and to support further research at the University of Wisconsin with the majority of license royalties received. In the 63 years since its founding, WARF has granted over $150 million to the university while building up a multimillion-dollar endowment from over 2,400 disclosures, yielding over 450 patents and 200 licenses on approximately 100 of the technologies. Total net income from patents was over $30 million from 1929 through 1985. The majority of WARF's income has come from investments that were created from the flow of license royalties on patents.

WRF. The Washington Research Foundation is a comparatively young organization, having started its operations in January of 1982. WRF was created to help the state of Washington's universities and research institutes protect and license the technology and expertise generated by their research. WRF also administers the Biological Materials Distribution Center, to make life-science materials developed by researchers at the University of Washington available to scientists in other universities and industrial research laboratories.

In just five years, WRF has generated income that has provided over half a million dollars to the University of Washington. By late 1987, WRF had evaluated over 250 technology disclosures and was administering 110, including nearly 40 license agreements. Nearly 98 percent of

WRF's disclosures have come from the University of Washington. The remaining 2 percent have come from the WTC. WRF also has a technology administration agreement with Washington State University, to serve its research foundation as a backup resource.

Important Factors for Successful Implementation. Six factors have been crucial to WARF's success in building a significant research fund. These are a successful initial technology and patent position that generated immediate royalties; substantial time devoted by skilled, independent alumni and vigorous efforts led by the initial foundation director; fortunate financial management, which avoided substantial losses in the 1929 stock market crash and onerous tax and antitrust interpretations; imaginative marketing, legal defense, and money management; a flow of significant technologies from university research results; and substantial contributions by those technologies to human health and welfare.

Experience at the University of Washington and at WRF seems to confirm the importance of several of these factors and can probably be generalized to other universities and agents. Most important is the flow of significant technologies, since that is the starting point for any technology-transfer program. It is important to have highly motivated, skilled people providing patent protection and marketing efforts on behalf of the technologies, whether inside or outside the technology-transfer office.

Of equal concern, but outside the direct control of the technology-transfer office or organization, is management oversight by higher administrators or directors. One key element that must be taken into account in such oversight is the length of time required for a technology-transfer program to succeed. The time needed is generally about ten years.

Initial evaluations of a disclosure typically take two to three months. Evaluations include literature and patentability searches, discussions with inventors about important aspects of the technology, development of a nonconfidential description of the technology for outside parties, and initial discussions with knowledgeable individuals in selected industries for further insights into applications or competing technologies.

When initial evaluations look promising, parallel efforts on patent protection and marketing are usually undertaken. The patent process typically takes from two to five years. Discussions with companies that may use the technology and be interested in licensing it can be held while the patenting efforts are under way. A year after the filing of a U.S. patent application, a decision must be made on whether to file patents in foreign countries. Foreign filings are expensive, and so this decision is especially difficult if a licensee has not been identified at that time.

Only a minority of disclosures will be licensed for commerical development, typically between 10 and 30 percent. Only a fraction of the licenses will generate significant royalties. WARF's experience is that only ten patents, out of more than four hundred obtained, have generated 90 percent of its royalty income.

Even when a license agreement is completed, only a small portion of the proceeds of a successful technology is paid upon signing as the up-front fee. The greater portion of proceeds comes from royalties based on a percentage of sales. This step follows the completion of product engineering, market introduction, and successful promotion by the licensing company, a process that takes two to three years for market introduction and another two to three years for sales to reach significant levels, if the program is successful. Hence, significant royalties may begin to flow about three years after the license is signed but are likely to be greatest seven to ten years after licensing.

Incentives and Monetary Distribution Policies. Regardless of the operating model chosen, technology-transfer programs are not likely to succeed without meaningful incentives for inventors. The type and range of incentives will differ from one institution to another, but one that will be common to all is the fee/royalty distribution formula. Most academic practitioners of technology transfer have concluded that the formula must be fairly generous toward the inventor and the inventor's department. Successful programs seem to allocate one-third to one-half of the fee/royalty income to the inventor. He or she must prepare and submit the invention disclosure and then help university administrators and licensing agents to understand the invention and its possible applications. Further assistance is typically required when the patent attorney is drafting the patent application or responding to actions from the U.S. Patent Office. Still more assistance is routinely required when the university or its agent begins to interact with prospective licensees, and this stage frequently continues even after a license has been consummated, in order to impart critical expertise from the inventor to the licensee. Generous incentives are required to encourage inventors to participate in technology-transfer programs and assist universities, patent attorneys, and licensees in the ways that are necessary to maximize the speed and ultimate success of the transfer.

Other incentives include the variety of suppoort services that can be offered by the university's technology-transfer office or by its agent; additional industrial funding, spawned by technology-transfer activities and increasingly recognized in academic reward systems; and favorable publicity related to new inventions, the issuance of patents, the execution of license agreements, and so on. Particularly in new programs, it is important for the institution to gain the active involvement and strong support of deans and department chairs. Unless they carry positive messages about the program to their faculty and actively encourage faculty participation, the program is not likely to succeed.

Directions for the Next Decade

We see continued increase in the number of university technology-transfer offices, as well as increases in activity, personnel, technologies, and economic return. The long time required for new offices to break even, and the importance of university-developed technology in boost regional economic development, are likely to lead to increased participation by state economic development departments in directly supporting technology-transfer offices and functions.

Liability. One area of concern that has not been clearly defined is whether liability suits for injuries suffered in relation to the manufacture or sale of products developed at universities will drag universities into lawsuits. Typical license agreements include indemnification clauses to protect the university, its officers, and its faculty from legal liability in any actions arising out of the exercise of the license agreement by the licensee. Nevertheless, the strength of these indemnification clauses has not yet been thoroughly tested in court.

The use of outside agents to handle some technologies adds another level of insulation between the university and legal liability. When the university uses an internal foundation to handle technology transfer, the degree of insulation may be less. Including management of the university's investment portfolio in the same internal foundation may be riskier yet. Until this issue is tested in court, it is not clear to what degree universities are liable for damage awards that can be assessed against them.

Strengthened Economic Development. The importance of university-developed technology to regional economic development will spur an increase in support of technology transfer by state and local government. This support may take the form of increased coordination with technology-transfer offices and agents, subsidies and grants, joint university-governmental development of incubators and research parks, and spurs to technology acquisition by local businesses.

One aspect that deserves increased attention is use of a developmental laboratory to move basic technology out of the university and develop working prototypes. This practice can increase the attractiveness of university-developed technologies and reduce the risk to industry of pursuing unsuccessful technologies. This concept is currently being explored by the city of Seattle and the University of Washington as a major focus of a research park/incubator facility under design by both groups.

While technology transfer is an important and exciting new frontier for academia, it cannot and should not replace or reduce the traditional roles of creation and dissemination of information by universities through research, teaching, and publication. Technology transfer should be undertaken in a manner that complements these basic goals of academic institutions. If done with that philosophy, and with an eye to the suggestions and caveats discussed in this chapter, it can be appreciated and valued by faculty, administrators, and friends and supporters of our universities.

References

Birch, D. "Interview—Why Things Aren't as Bad as They Seem." *Inc.*, 1981, *3* (9), 41-44.

"Corporations on Campus." *Science*, 1987, *237*, 353-354.

"The Economy of the 1990s: Where to Live—And Prosper." *Fortune*, Feb. 2, 1987, pp. 52-56.

General Motors v. *Devex*, 461 U.S. 648, 217 USPQ 1185 (1983).

"Growing Places." *Inc.*, 1986, *8* (10), 57-66.

Hammond, G. S. "The Academic-Industrial Interface." In N. S. Kapany (ed.), *Innovation, Entrepreneurship, and the University*. Santa Cruz: Center for Innovation and Entrepreneurial Development, University of California, 1978.

Jonas, N. "Can America Compete?" *Business Week*, Apr. 20, 1987, pp. 45-47.

Ramo, S. "Technological Innovation: A Social-Political-Economic Problem." In N. S. Kapany (ed.), *Innovation, Entrepreneurship, and the University*. Santa Cruz: Center for Innovation and Entrepreneurial Development, University of California, 1978.

U.S. General Accounting Office. *Patent Policy—Recent Changes in Federal Law Considered Beneficial*. Washington, D.C.: U.S. General Accounting Office, 1987.

Whipple, R. P. "A New Era in Licensing." *Les Nouvelles*, 1987, *22* (3), 109-110.

Gerald A. Erickson is past president of the Washington Research Foundation. He is cofounder and vice-president, chief operations officer, for Vehicle Optix Company.

Donald R. Baldwin is assistant provost for research and director of technology transfer at the University of Washington.

As American businesses streamline their operations to meet the challenge of competitiveness, universities are being called on to play a strong facilitative role.

Higher Education in Corporate Readaptation

Ronald R. Sims

In the past decade, American corporations have found themselves trapped in the worst of all possible worlds. Although profits still seemed strong, by the mid-1970s productivity growth was slowing to a crawl. By the late 1970s, the competitiveness of American manufacturers, as measured by their shares of world markets, was sagging. The Federal Reserve Board pushed up interest rates to fight inflation, and the dollar soared. American corporations found themselves having to cope with higher credit costs, while being priced out of overseas markets and surrendering big chunks of their domestic markets to cheaper and better imports.

Increasingly in today's economy, American firms need help to compete in a highly competitive, technologically advanced, and rapidly changing global economy. Universities, with their knowledge-based resources, can help corporations adapt to their expanding roles and to the challenges of a new economic order. Some campuses have become involved in providing services to businesses, including economic policy analysis, applied research, capacity building for economic revitalization, human resource and new business development, and stimulation of technology transfer, to name a few. University involvement in corporate streamlining represents the proactive and strategic role higher education

J. T. Kenny (ed.). *Research Administration and Technology Transfer.*
New Directions for Higher Education, no. 63. San Francisco: Jossey-Bass, Fall 1988.

must undertake to help our nation address declining productivity and challenges from foreign competition.

In the following analysis, we will discuss, by example, the concept of corporate readaptation, describe how universities can help businesses in the readaptation process, present a recent case of a successful university-corporate relationship, detail some potential problems facing research managers who represent universities in corporate partnerships, and highlight some recommendations that higher education should follow in developing policies to increase the potential success of the university-corporate relationship.

Corporate Readaptation

The adjustment to the 1970s has been painful for American firms and for the United States in general. Plant closings, layoffs, restructuring, renewal, readaptation, mergers, and acquisitions have been the main focus of most corporate activity during the 1980s. In addition, some industries have undergone wholesale elimination of excess capacity in a process of consolidation designed to carve up a shrinking pie among fewer companies.

The U.S. steel industry has slashed employment by 44 percent, to 163,000 jobs, and reduced raw steelmaking capacity by 27 percent, to 112 million tons annually. In the paper industry, Medford, Pacific Lumber, Crown Zellerbach, Diamond International, Hammermill, and St. Regis are the biggest among a host of corporations that have disappeared in mergers or acquisitions. Consider Wells Fargo's acquisition of Crocker Bank, Chevron's purchase of Gulf, or Ryder's investment of $500 million in aviation since the 1980 deregulation of trucking ("America's Leanest and Meanest," 1987). United Airlines, AT&T, IBM, EXXON, Union Carbide, and even the major television networks have been "slimming down" in an effort to rise to the challenge of tougher competition (Kenny, forthcoming). The overall goal of American business has become cost reduction and the improvement of long-term profitability and growth.

It is not often that one can open a newspaper, listen to the news, or read a magazine without learning about another new bankruptcy, a hostile takeover of a once proud company, or relentless currents of losses from companies that seemed invulnerable only a few years ago. American corporations have a long way to go to reclaim the markets they have lost. Unless they become more efficient, a significant decline in the American standard of living may be unavoidable. To address these negative prospects, many industries are starting to renew themselves by going through a readaptation process. Just what is this process?

Lawrence and Dyer (1983) introduced the concept of readaptation, which is the organizational state of being simultaneously efficient and innovative. This is an interesting approach to renewal, because efficiency and innovation are in conflict with each other. Efficiency requires careful use of resources and often leads to a highly specialized and bureaucratic structure. Innovation is concerned with being on the leading edge of offering the new products and services that are desired by the environment. Efficiency and innovation are difficult to reconcile in the same organization, because they require different skills. Cost cutting may inhibit innovation; innovation may require a loosening of internal control.

The point of readaptation is that today's and tomorrow's successful corporations must manage to reconcile these two forces. To sustain the readaptation effort, they must learn to be efficient and innovative. Maintaining an influx of fresh ideas while still being efficient requires continuous trial and error and interpretation of competitors and of the environment. Companies must also treat information as a competitive advantage, use the flexibility as a main strategic weapon, and encourage learning and adjustment by all corporate members. Striving for both innovation and efficiency energizes the corporation into a learning mode that enables it both to avoid the pitfalls of and to respond to bloated management structures and a business-as-usual attitude. Efficiency and innovation have become mechanisms through which firms can be activated toward both learning and readaptation. Successful corporate readaptation keeps the organization healthy, flexible, and vital.

The corporation's role in readaptation is very direct: Management must operate efficiently to realize adequate returns on investments. The unique entrepreneurial function, however, is to innovate: to develop new products and processes, invest in new human resource development programs and capital facilities to produce new products, and implement cost-reducing technologies. For many corporations, this function involves substantial investments in research and development, an alert purchasing policy that takes advantage of new equipment or other producers' goods, and corporate changes expected to reduce costs and increase productivity. It also means providing training and retraining of workers to operate the new plants and equipment and helping workers find new employment when readaptation calls for the streamlining or closing of plants.

Many U.S. corporations have started to develop a tough approach to doing business by cutting costs to the bone, selling marginal businesses, closing inefficient plants, and laying off personnel. They have also gone much farther to boost productivity by becoming innovators in product development, marketing, and distribution and by constantly discovering ways to do things better and more cheaply.

Georgia-Pacific, for example, to achieve greater efficiency, patiently sold unprofitable divisions and moved into paper businesses with higher profit margins. At some companies, such as General Motors, new

techniques have stirred up a hornet's nest of problems, including production delays and higher overhead costs from balky machinery. Other
companies, including Crown Cork & Seal, have evinced great success.
Crown has raised labor productivity by 50 percent in the past five years,
partly by shutting down its older can-manufacturing plants and moving
production to newer, more automated factories with twice the capacity of
the older ones ("America's Leanest and Meanest," 1987). To improve
productivity, other companies have brought in battalions of robots or
built overseas factories in low-wage countries.

Still other firms have moved toward greater innovation. Little
Marion Laboratories, Inc., successfully navigated its way through a flo-
tilla of larger rivals by licensing pharmaceuticals from foreign companies, rather than spending billions to develop its own drugs. Innovation
can be as simple as the Walt Disney organization's creation of a separate
film company, Touchstone Pictures, so that it could produce films for
a more adult audience, without losing its immensely profitable family
image ("The Renewal Factor," 1987). Innovation can also be dramatic,
like the Squibb Corporation's delving into the discovery of wonder drugs
for the 1990s.

Higher Education's Role in Readaptation

The American Association of State Colleges and Universities
(AASCU) (1986) recently looked at emerging roles for public colleges and
universities in a changing economy. Through a series of case studies,
AASCU was able to identify a number of areas ripe for cooperative
activity. These were human resource development, economic and policy
analysis and research, capacity building for economic development,
research to develop new knowledge, technology transfer, and support for
the development of new knowledge-based businesses.

While the role of the university in corporate readaptation is not
so immediate as that of the corporation, it is fundamental in that the
educational system is largely responsible for transmitting and enhancing
the body of knowledge through the generations. It is mainly through
educational processes that nonmaterial investments in human resources
are made, which prepare people for productive roles in the economy and
in the broader society. Moreover, increased human resource investments
are considered a major element of increasing the potential success of
corporate readaptation and productivity.

Institutions of higher education prepare the scientists, engineers,
business administrators, and other actors who create innovations, develop
them for commercial use, identify more efficient ways of operating, and
make decisions to invest in new processes, plants, equipment, structures,
and cultures. A coordinate function of academe is to perform basic and

social research that advances knowledge, which feeds into applied research and development and corporate efficiency and in turn benefits from them.

More broadly, higher education as a whole educates and prepares future members of corporations to operate the increasingly complex technology of their organizations and assists in specific training through special courses or on-the-job instruction. The educational system also contributes indirectly to human resource mobility, because more highly educated workers are generally more mobile. While the efficiency and effort with which today's workers perform their duties are responsibilities of management, the values and attitudes of those workers are an important background element that can be influenced by higher education. Through its general understanding of the economic system and of the role of productivity and growth in the realization of individual, corporate, and social goals, higher education can be an effective participant in readaptation and economic revitalization.

Through a variety of social and economic analyses and projections of trends, university research can contribute background studies required by corporations for long-range planning. During readaptation, such planning facilitates resource reallocation in response to anticipated shifts in supply and demand and to other dynamic economic and social forces. Serious deficiencies of information can be obtained only through corporate experience, which may have a large element of unpredictability. Other information can be provided through well-researched, analytically sound, and objective studies on a variety of corporate problems. Campuses and corporations must seek common ground with respect to the roles and functions that higher education can perform during readaptation.

Research and service units can act as troubleshooters when an industry is facing some special difficulty, or as advisers helping an industry evaluate its own projects or ideas. In addition, they can serve as gatekeepers to a wider scientific and business community, brokering the many kinds of research and information available through a campus's different departments. The most important role that higher education can play in readaptation, however, goes beyond these immediate advantages. Through their research and development capabilities, universities can assist in acquiring knowledge about the needs of various industries and can therefore identify the specific ways in which such needs can be met.

Serious deficiencies of information often impede organizations' ability to meet readaptation goals. Such goals may include developing new products and services, identifying customers' needs, and meeting the challenges of competitors. Universities can help corporations gather, analyze, and effectively use various types of information. The most obvious example of this function is in research. Campuses can offer a broad spectrum of assistance, from pure research to research aimed at specific products or human resource development. One example is the ongoing relationship between Monsanto and Washington University–St. Louis in biotechnology research and other areas. Monsanto also sends twenty of its best chemical engineers to the university for an intensive yearlong update on current developments in the field, ranging from computer applications to problem-solving techniques. More than 90 percent of these engineers complete the program and go on to better positions in the company. Thus the program ensures that a new cadre of personnel will continue innovations.

The same idea drives the concept of lifelong cooperative education, developed in 1982 by four professors at Massachusetts Institute of Technology (MIT). MIT helps high-technology companies keep their engineers up to date on state-of-the-art developments through tutored video instruction. MIT has successfully assisted executives and other members of such corporations as IBM, Bell Labs, and Honeywell Information Systems. Stanford University has done the same type of work with engineers at Hewlett-Packard. Cornell University is also active in helping corporations, such as IBM and Floating Point Systems, to meet today's and tomorrow's economic challenges.

Cooperative research—like the kind performed at the Microelectronics Innovation and Computer Research Center in the Research Triangle Park (it draws on the research staff from North Carolina University and Duke University), at the Stanford Center for Integrated Systems (sponsored by seventeen microelectronics firms, including Hewlett-Packard, Intel, and Xerox) and at MIT's Microsystems Industrial Group (consisting of large and small companies)—exemplifies current higher education–corporate partnerships. Whatever a firm's readaptation policy may be, an essential element of its evolution and implementation will be the availability, training, and development of its human resources. Applied research units can assess such needs in a systematic way, package courses of discovered relevance, and supply them through various forms. Very often, the responsibility of providing competently trained personnel falls almost entirely on higher education.

For many years, higher education has provided quality instruction to workers and, during readaptation, now plays a more significant role in their continued training and development. The quality of the education and training that can be provided directly affects the quality of those responsible for readaptation. Readaptation addresses a variety of educational and training-related needs, including accommodation of changes that result from reorganization, streamlining, mergers, or acquisitions; preparation for changes in the skills and knowledge that will be required by a corporation and its personnel; and improvement of the skills and performance of people in their current jobs.

Campus-based research and service units can design programs that focus on the acquisition of technical, manual, and interpersonal skills or skills for self-awareness, career development, and so on. Such programs may take the form of on-site or off-the-job training. As businesses restructure themselves in an increasingly complex economy, the knowledge and skills required of tomorrow's workers will continue to change at a rapid pace, and there will be a corollary upgrading of the minimal skills required. To be competitive, corporations need to facilitate the kinds of knowledge transfer that will enable them to remain on the cutting edge of new technology. University faculty are a prime source of new knowledge and techniques to assist in the development of workers.

Readaptation, as a worldwide phenomenon, has been on the increase in recent years and is likely to continue growing for some time. No industrialized or developing society can afford not to learn about and deal with the impact of change on individuals, families, and the corporation itself, together with its interdependent environments. The need for universities to apply their investigative, interpretative, and educational skills to studying the effects of expanded readaptation is extremely crucial, for humanitarian, social, and organizational reasons. Involvement in the process provides an important opportunity for universities to make significant contributions that can affect business practices now and in the future.

A University-Corporate Partnership

Auburn University at Montgomery's (AUM) relationship with the Monsanto Agricultural Company (Anniston, Alabama) represents the kind of role a university might play in a corporate partnership. AUM's Center for Business and Economic Development (CBED) worked with the management of Monsanto to develop a quality-of-worklife (QWL) program. For the purposes of this project, workshops were designed in which CBED facilitators assisted the company's top three levels of managers in identifying, diagnosing, and setting priorities for problems and in developing plans of action. The plans were supposed to result in operational programs for improved productivity in the plant, increased participation by employees in the management of the plant, and increased morale and work satisfaction among employees.

After initial negotiations between the CBED director and Monsanto's top managers, AUM entered into a contract with the company. The first stage of the program called for the CBED staff to interview each company manager and gather data on current working relationships among these executives, their expectations of a QWL program, and their general impressions of the current effectiveness of Monsanto's organizational structure. The questionnaire used in the interviews was specifically designed for Monsanto. The CBED team compiled and analyzed the information gathered during the interviews and then met with Monsanto's managers to discuss their findings. This discussion framed the specific issues that were important both for developing a QWL program at Monsanto and for defining the role of CBED. The issues to be addressed, and the plan of action for implementing principles of team building and organizational development, were thus established jointly.

During the second phase of the project, the CBED staff took Monsanto's plant manager and superintendents to a two-day, off-site retreat at a nearby conference center. The purpose of this retreat was to articulate a cohesive management philosophy and to establish a definitive strategic direction for the company. The goals, objectives, and plans for implementing the philosophy and for guiding the strategic direction of the company were written for each division represented by the company's leaders. In addition, some issues concerning interpersonal relations and conflict resolution were discussed at the retreat.

Eight weeks later, the CBED staff conducted a second off-site retreat, for a day and a half, with second-level managers (supervisors). Four weeks after the second retreat, a third off-site retreat was held with third-level managers (foremen and others). Before each retreat, the same interview process that had been used with Monsanto's executive team was also used with the other groups. At the first retreat, the CBED staff and the executives agreed that later sessions would include half a day for each group to meet with the next-highest management group and share concerns. CBED facilitators were at every meeting and met weekly with executives.

After each retreat, team leaders visited the plant to help managers and foremen follow up on issues covered at the retreats, met with employees throughout the facility, and coordinated the QWL program for the company with the plant's union representatives. This process called for interviews with union leaders, brainstorming sessions with them on union-management relations, and attendance at union-management meetings.

After these initial phases, the team leaders continued to visit the plant almost every week to assist the company. The CBED director also talked by telephone with the plant manager and visited the facility as necessary. Therefore, the first efforts were not lost in a haze of good intentions; there was a continuous effort made by the university, through CBED, to respond to the company's needs and to follow through with the program that had been mutually established.

The partnership benefited the campus by giving its units valuable experience in working with a company during readaptation on a long-term, complicated, ambitious program. This partnership also provided material that was used later in teaching classes on organizational behavior and management.

At Monsanto, the project directly affected the whole plant during readaptation. Specifically, the QWL intervention allowed the company's management to do the following:

- Bring issues and problems into the open
- Determine the magnitude and potency of problems
- Provide the plant manager and his staff with an assessment of the lower managers' attitudes and concerns
- Collect information on employees' needs, problems, and frustrations in an organized manner, so that corrective actions could be taken
- Understand subordinates' feelings and attitudes and investigate better methods of communication in specialized areas
- Make clear to everyone in the company that top management was attempting to improve working relationships
- Establish mechanisms by which all members of each management group could feel that their individual needs were being noticed
- Provide additional mechanisms for supervisors and foremen to influence the whole organization
- Establish operational programs for improving productivity and encouraging innovations in the plant
- Evaluate the current organizational structure and design a more efficient one
- Articulate a cohesive management philosophy
- Develop a realistic and guiding strategic direction.

In addition to the Monsanto project, CBED and other AUM campus units and departments have become involved in partnerships with other corporations and government agencies, providing professional training and development programs on a variety of management issues. The campus has seen positive results from its response to the increasing need for knowledge-based services among businesses in its region.

Potential Problems

The university-corporate partnership is not without its problems, and a project manager must be aware of, identify, and respond to them on a regular, flexible basis if the campus is to remain a constructive·and useful partner. During readaptation, the corporation requires multiple forms of assistance, which is often impossible to obtain from individual academic departments and which requires the cooperation of many campus units. The project manager must bring about effective collaboration between relevant fields and departments. Such collaboration in turn requires the formulation of a program concept that can intellectually mobilize the resources that are dispersed through various departments.

With appropriate administrative and departmental support, the project leader can be a versatile and productive resource to the corporation. In principle, higher education can deal with almost any type of problem encountered during the process, from basic research (technological development), routine consultation (most appropriate corporate structure), and policy development (modification of corporate rules and regulations). Nevertheless, the particular strength of a university's program will be determined by how well the project manager develops means of matching the needs of the firm to the university's capabilities. (One form of mismatch occurs when a firm is led to believe that it can obtain services that the university is not competent to offer.) The manager must also ensure proper integration of the contributions made by the departments and be able to handle potential conflicts over priorities, demands for secrecy, incommensurability of values, working rhythms, departmental fragmentation, and so on. These are endemic problems, and there are no simple solutions.

Suggestions for Successful Partnerships

Wilson (1980) has discussed several elements of successful partnerships. To structure an effective and sustaining relationship, the parties must first cooperate. The outcomes of the partnership should be clear, as must the benefits to be derived by both parties. There also should be an ongoing assessment of the products of the relationship, as well as an assessment of the relationship itself. Relationships must evolve through an increased understanding of the mission, goals, and relevant issues of both organizations. The following recommendations are offered:

1. Understand the private sector's expectations and needs. Listen to what the corporate representatives are saying, and react accordingly. During readaptation, corporations do not appreciate "canned" approaches and theoretical programs. The applications of any intervention, consulting, training, or service provided by the university must be made clear. Because this is often not done, corporations sometimes have little faith in academic consultants and university services.

2. Analyze the nature of each corporation. An institution's track record in university-corporate relations is important, but once the particular program has begun, the corporation wants to be treated as a unique entity.

3. Recognize that corporate managers appreciate specificity, and they expect results that can be measured by some reasonable method.

4. Stay in touch with the principal managers at the corporation; do not allow a lull to interrupt the relationship. Too often, managers feel that consultants call them only when they are paid to do so. In a university-corporate partnership, this is not acceptable behavior.

5. University leaders must advocate a proactive role in evaluating current and future policies for recognizing and taking advantage of the new opportunities that exist in the partnership. The university must be flexible in its efforts to meet traditional and future missions. It must develop policies that allow an expanded ability to respond to corporate needs and must encourage faculty members to use their expertise, without detracting from their academic duties.

6. The university must develop a mission that articulates a philosophy of strategic involvement in the problems of a rapidly changing global economy. Whether in the form of a service or of economic development, the mission must recognize and accept the challenge to work more closely with the private sector. An institution must build a culture that supports expanded partnerships.

An important key to successful integration is the balancing of the two organizations' interests. To foster the partnership successfully, both organizations must develop a strategy for making the relationship work, rather than just letting it happen or waiting until foreign competition or decreased productivity become insurmountable problems. The resulting benefits for both organizations may be limitless.

Thoughts for the Future

The accelerating rate of change is producing a business world in which customary managerial habits are increasingly inadequate. Experience was a useful guide when changes could be made in small increments, but intuitive and experience-based management philosophies are inappropriate when decisions must be strategic and have irreversible consequences. In light of the discontinuous, large-scale changes facing the world, systemic modification becomes inevitable and reflects the evolution of products, services, markets, organizational structure, culture, and human resources.

Allan Kennedy, coauthor of *Corporate Cultures*, describes the direction in which today's corporations are headed: "If I had to go 'way out on a limb, I would say that large corporations, as we have known them will not exist in thirty years' time" (quoted in Naisbitt and Aburdene, 1985). Kennedy says companies will be unable to retain their people, because too many others will have gone out on their own and struck it rich. Instead of in thirty years, Naisbitt and Aburdene (1985) think Kennedy's prediction will come true in half that time, by the year 2000. In either case, adaptation to radical change will be an ongoing concern.

Research universities have a special ability and responsibility to contribute to the American corporate initiative, in both the short and the long term. The role of campuses in readaptation will be to assist the private sector in defining and following an innovative political, social, and economic trajectory within a wide range of possibilities. Campuses have an indispensable part to play in defining feasible solutions that may lead to improved productivity and growth. Basic and applied research, along with training, enhances the corporation's store of knowledge, and this will ultimately help to ensure high levels of innovation and lay the groundwork for a competitive response to the challenges of tomorrow.

References

American Association of State Colleges and Universities. *The Higher Education–Economic Development Connection: Emerging Roles for Public Colleges and Universities in a Changing Economy.* Washington, D.C.: American Association of State Colleges and Universities, 1986.

"America's Leanest and Meanest." *Business Week,* Oct. 5, 1987, p. 80.

Kenny, J. T. "Advocacy for University Research and Service Units." *Journal of the Society of Research Administrators,* forthcoming.

Lawrence, P. R., and Dyer, D. *Renewing American Industry.* New York: Free Press, 1983.

Naisbitt, J., and Aburdene, P. *Reinventing the Corporation.* New York: Warner Books, 1985.

"The Renewal Factor." *Business Week,* Sept. 14, 1987, p. 102.

Wilson, J. *Models for Collaboration: Developing Work-Education Ties.* Washington, D.C.: American Society for Training and Development, 1980.

Ronald R. Sims is an associate professor in the School of Business Administration at the College of William and Mary.

How can universities develop the resources to assist units
of government in meeting a new generation of challenges?

The University in Service to State and Local Government

James T. Kenny

State agencies and a number of larger units of local government are coming to require increasing amounts of technical service, systems-management expertise, training, and organizational assessment as they strive to meet their mission requirements in an expanding service environment. Specific units often lack organizational development plans and frequently see themselves as unable to make constructive changes in an age of fiscal austerity and diminished levels of federal support. The irony is that such nonfederal entities are asked to shoulder a greater share of administrative and social responsibilities in today's society. In effect, they are called on to do more with less in a growing service arena. Some universities are responding to the call for help with new, cost-effective partnership initiatives. Such efforts are preparing agencies to meet their technical, training, and even legal problems in today's government milieu. In the process, a number of campuses are strengthening their public services and gaining a new kind of recognition.

An expanded state-government partnership gives a university an opportunity to bring its needed expertise to bear on the solution of a wide range of administrative and technical problems. A university may be an agent of modernization and change, but a campus that wants to be of assistance may find that it, too, needs to redefine or upgrade some of

J. T. Kenny (ed.). *Research Administration and Technology Transfer.*
New Directions for Higher Education, no. 63. San Francisco: Jossey-Bass, Fall 1988.

its traditional views of service and its sense of mission. What seems hopeful in this changing environment is the promise of a redefinition of the historic service relationship and, for forward-looking campuses, a new sense of direction and relevance in tomorrow's information society.

Organizational Development

Most modern universities, even smaller ones, provide an array of academic specialties of demonstrated use to governmental entities seeking an upgrading of their administrative services. Campuses may and often do offer flexibility scheduled credit courses in such traditional areas of interest to public managers as public administration, industrial psychology, accounting principles, and management. Continuing education and short courses are often tailored to career-conscious public servants, and special efforts are sometimes made to secure state or local agency contracts for blocks of skills-upgrading services for public-sector employees. Workshops and seminars are used to good effect to bring university faculty face to face with government workers in such specific skill areas as personnel management.

Specialized comprehensive programs, such as programs in certified public management, may also be considered as an institution's service capacity grows. In this case, a specialized host unit is usually assigned the task of coordination. This may be a center for governmental services, with staffing and resources at its disposal to begin the work in a credible fashion. This author has firsthand knowledge of the management training program at a midsize state university. This institution's Center for Government and Public Affairs, as it began its program of service, offered specific short-term services to individual agencies. This was accomplished through the coordination and brokerage of faculty-supplied expertise drawn from cooperating academic units. Specific cost-reimbursable contracts were entered into with units of state and local government, to provide solutions to such problems as tax revenue forecasting, improvements in promotion-selection procedures, training in performance planning and appraisal, policy review, job classification, and pay-scale revision. These small-scale projects, which were labor intensive and carefully performed in a cost-effective manner, helped establish the reliability of the university, its faculty, and its specialized units.

The good public relations, and the direct and indirect cost recovery resulting from these initial contracts, led campus officials to the conclusion that specific university services to state and local government not only constituted a logical and high-profile extension of the university's service mission but also could generate welcome revenue. A next strategic step was the securing of budgetary line-item funding for permanent staffing of the campus service program—a step that, by then, was encouraged by state agencies and the governor's office.

Full-time staffing enabled the university and the center to manage the flow of work, engage in long-range planning, and seek additional contracts. With additional resources available, an instructional component was soon added, enabling the unit to provide facilities and staffing for an increased number of both on-campus and on-site classes for state employees.

With the structure thus created, the institution was able, with some lobbying, to secure legislative appropriation for a certified public manager (CPM) program. The intent was to provide a training vehicle for agency-identified full-time employees who might be upwardly mobile or ready to move into supervisory positions. The program, as developed, was not unique to the host campus, and many good ideas were solicited and adapted from member organizations of the National Certified Public Manager Consortium. In developing the model, some natural in-system allies, such as the State Employees Association and the State Personnel Department, were helpful. Programs geared toward the improvement of performance, technical competence, and skills of state employees are usually supported by these entities. Legislative and executive-office support was also ensured with the promise of a cost-effective plan to upgrade the state's work force. The program was represented as a comprehensive management development and training program, to be offered by the university but to be conducted under the auspices of the State Personnel Department. An advisory board was created, including members of the State Personnel Board and the governor's staff, participating agencies of government, and university officials. In its design, the model closely paralleled the better-known certified public accountant (CPA) program in that it, too, was based on coursework, job-related applications of instruction, examinations, and professional recognition through certification.

The university's CPM program consisted of six levels of training. This included eight weeks of classroom participation in core courses of state government management. Specific levels included instruction in organization of state government, organizational communication, time management, management information systems, organizational management (including fiscal management), and program evaluation. Each core course was modified to meet the unique situation. Only highly qualified faculty were employed, and some of these were drawn from other institutions of higher education and from the professions. Selected training modules included videotaped and media-assisted training, case studies, small-group seminars, simulations, and lectures. While courses were initially held only in the state capital, but its second year of operation the program was being offered on-site to employees in each of the state's administrative regions. This program, once in operation, helped make the host campus's name synonymous with high-quality education and cost-conscious service in the halls of state government.

Technical Competence

Training in technical competence suggests yet another approach to a strengthened university-government partnership. Public agencies are being challenged to stretch tax revenue dollars and seek new efficiencies in administration. Data-systems management and expanded computer applications are eagerly sought in all phases of state operations. Here, too, universities have an advantage. Many public officials, recognizing the need for improved records management, are wary of hardware vendors who are eager to sell but have little sympathy for systems integration, expanded applications, problems of duplication, and limits to the cost of equipment acquisition. This offers an opportunity for a campus service center to provide comprehensive data-systems management training for state supervisory personnel.

Training of this nature is costly. States and cities are aware of this and are generally unwilling to bear the costs of wholesale in-house staff upgrading. Data-systems management training is usually conducted on an ad hoc basis within each agency, or in some agencies but not in others. This leads to uneven levels of proficiency unit to unit, diverse systems orientations, and a proclivity toward the acquisition of exotic and sometimes incompatible hardware. Many units are unwilling to recruit training professionals, even when such staff may help increase organizational efficiency. This is a point worth remembering in all areas of governmental service provision. It is often much less costly for an agency to contract a service, rather than seek an in-house solution that involves the creation of new line items that might invite intense public and legislative scrutiny. A short-term and shortsighted solution to the training problem is the sending of state employees around the country to expensive and sometimes irrelevant technical training programs. Elected officials and the media will also be inclined to target such travel as an unnecessary burden on state or municipal resources.

A local university, then, does offer some real advantages. For one thing, training expenses may be kept down to cost recovery plus indirect costs. For another, the necessary expertise is there for the taking. Start-up costs constitute a problem, for it will be necessary for a campus service center to have a classroom equipped with dedicated devices and terminals for trainees' use. Such devices will need interfaces with participating state agencies' systems, to provide as realistic and relevant a training program as possible. Therefore, supervision during access periods and coded security becomes a necessity. A university may purchase the requisite hardware, seek a bilateral agreement for its acquisition, or secure the needed equipment under the terms of a multiagency training agreement. Whoever buys the machines, the arrangement offers such long-term savings that the initial outlays are eventually recovered. This is especially true if a multiagency agreement can be put into place, ensuring regular daily use of the space and training hardware over the contract period.

A government service center pursuing this kind of long-term training agreement will want to seek the assistance of faculty experts, the university computer services center, and perhaps the learning resource center in preparing a prospectus for agency review and consideration. As in all proposed major agreements, interagency discussion should, after preparation, begin on campus (preferably in the president's dining room), with a fully briefed executive hosting top agency officials and commissioners. Training in technical competence has certain inherent advantages for an institution of higher learning. It, like a CPM program, gives a campus needed visibility with officials and lawmakers, offers some possibility for a long-term and lucrative partnership, and encourages the university's continued upgrading of its commitment to data automation and state-of-the-art training in this field.

Econometrics

In most states, the governor's office maintains a close working relationship with its department of finance. This unit, with its taxation and revenue divisions, is faced every year with the onerous task of estimating state revenues from all sources and then putting such information to use in the construction of the governor's budget. The fortunes of more than one state's finance chief have declined when inaccurate revenue estimates resulted in budgets far in excess of state revenues. Finance offices need to get an accurate view of revenues and specifically require some reliable measurement, with predictable utility, to estimate tax income and other revenue sources in a given fiscal cycle. Many states employ budget analysts and tax experts, who try to look at the multivariate dimensions of revenue projection. Sometimes their best estimates are close; sometimes they are not.

National-scale economic forecasting is readily available by subscription through Chase Econometrics Services and other entities, but the megapicture does not provide reliable local data to improve the estimates of state budget planners, and it lacks the local industry-specific barometers essential to accurate prediction. States need to, but sometimes do not, develop econometric models of their local systems. Econometric models vary in systemic design. Whatever model is preferred, a state-specific study needs to be completed, an integration with national data must be effected, and the system's designers need to run historical state data to validate the accuracy of the model. Some universities are interested in this type of work and are willing to engage in salesmanship to secure contracts for these services with state finance offices. One of the most accurate forecasting systems this author has seen was completed in under than two years. One economist, several graduate students, and a university computer were involved in the project.

Econometric modeling is not limited to state revenue forecasting. A good model can be used to analyze specific components of economic activity and, as the data are increasingly refined and disaggregated, can provide reliable predictive indicators for specific sectors. The garnering of such information will also be useful in a university's initiatives for services geared to business and economic development.

Litigation Support

Class-action suits absorb the time and resources of states, counties, and municipalities. Legal fees can be costly, and a government unit may find itself in the midst of a major claim or claims at almost any time. Agencies need the expertise universities can provide, not as a substitute for good legal advice but as an adjunct to counsel.

A university center will want to broker the expertise of a diverse faculty to provide a number of options. These may include informational services, which could help a governmental unit to avoid potentially litigious situations; expert witnessing, to enable the agency to successfully defend itself in court; or postdisposition support, to assist the state in meeting the terms of a settlement, decision, or consent decree. Many universities and individual faculty members are engaged in such services today. Class-action judicial settlements at both the local and the federal level, while often regrettable and tragic, have become common in society's scheme of conflict resolution.

Most class-action suits involve substantial documentation. Thousands of separate pieces of information, including depositions, need to be quickly assembled and just as quickly retrieved, beginning at the period of discovery through the time of appellate proceedings. The filing of such records in a usable fashion presents a storage problem for legal firms, since data management is not their particular forte. Contracting for the development of data-automated record and document storage is often a desirable alternative. Federal agencies now issue requests for proposals seeking contractors to develop data-management systems for specific classes of litigation. Some universities, even those with limited computing facilities, have been active in the pursuit of these contract opportunities, as have a number of small businesses. Major litigation-support opportunities may also be found with larger municipalities. At both the state and the local level, however, one will want to stay in close contact with procurement offices, because requests for proposals and bidding practices vary.

New Research Opportunities

Land-grant universities have enjoyed traditional relationships with state and federal units of government, which are sometimes fixed by law and usually supported by line-item research budgets. In addition to cooperative extension and agricultural experiment stations, such large comprehensive institutions are often involved in sea-grant programs, highway and water-resources research, labor relations, forestry projects, and other traditional service programs. Their applied research and assistance tend to be of high quality and need not be duplicated by other institutions. In an age of growing specialization, however, there are areas of short-term cooperation that non–land-grant campuses may want to match with their own academic specialities and organized units.

Organizational assessment may be conducted for government units that need structural or personnel improvements. Valuable surveys may be contrated to help regulatory agencies understand their policy outcomes. Demographic studies are in great demand by education departments and social service providers. Publication assistance is sometimes sought by agencies that provide public information, and model programs structured by a campus are sometimes a cost-effective way of testing and providing preliminary evaluations of new service-delivery modes. During the past year, this author worked with a campus rehabilitation unit that was conducting a mobility study among blind computer programmers. The results of this research were expected to help state vocational trainers identify visually impaired students with potential for success in this field. The same center was involved in a major dependent-care database assembly project. (Incidentally, the unit was being well compensated for its efforts in each of these areas.) Valuable work is going on at all levels. The university that does not see an expanded role for itself in today's information-hungry environment, or that refuses to make public-sector links, makes a tragic mistake of omission and shortsightedness.

State Funding Strategies

Any university center or unit with an abiding interest in state contracting and service extension into the public sector will want to work closely with its own governmental liaison office. An institution that has authority to conduct lobbying activities (and this varies from state to state) will have at least one point of contact in the president's office or in a unit that reports to the president. An institution periodically needs to bring together those in the ranks who work closely with agencies. The campus office of sponsored research should be involved, since research administrators have regular dealings with public-sector counterparts in grant and contract administration. Units with extant agreements should also be included in any dialogue aimed at expanding service contracts. State government is a world unto itself, with a set of behavioral norms, rules, and personalities that must be understood before cooperation is possible. Regular contacts must be kept on many levels. The key to capitalizing on opportunities lies in the garnering of reliable information. Frank exchanges among those campus officials most closely involved in ongoing service certainly helps. Many service-oriented campuses regularly invite state officials into the halls of academe for presentations, seminars, and luncheons. Such officials usually make willing advisory-board members for service units, and this also helps in network development. Legislators should be involved as well. A well-positioned campus will regularly invite its local delegation and long-term friends of higher education to such special events as the opening of a new research center, unit, or program. Legislative support for any university endeavor should be rewarded with some appropriate recognition on campus.

Staff development also plays an important role. Today one often finds highly qualified, educated, middle-level management people in governmental units. Whenever possible, a campus should offer adjunct faculty status to the best of these. What they have to offer our undergraduate and graduate students is a storehouse of practical in-the-field knowledge; what they have to offer the university in future opportunities is tremendous. When the opportunity presents itself, campus centers should recruit some of their full-time professionals from these agencies. Some of the most successful directors of research and service centers are people of academic repute who have worked in government. Professionals of this ilk generally have no difficulty securing contracts and grants to support their center's missions.

Lobbying, both in regular and in special sessions of the legislature, has certain advantages for state-supported institutions. The focal point of regular lobbying is the securing of general appropriations or bond issues for a referendum ballot, but some institutions successfully lobby for funding for specific long-term items that drive centers for research and public service. Attention needs to be focused on the appropriate legislative committees in both houses. The financial committees in each house (ways and means, finance, taxation) will be important if one is looking for inclusion of such a line in the state education budget. A sponsor will be needed to amend the budget bill and, of course, such a legislator should be a longtime friend of the university. Both timing and cultivation are important here; let the amateur beware.

Reliable and up-to-date information is of great importance to a legislator, both in making effective decisions and in securing alignments on important issues. Here, the same philosophy that drives our relations with state agencies may be profitably employed. Sharing of information, the basic commodity that educators bring to experience, may significantly enhance a university's legislative program. An institution must always be prepared to offer the members of its local delegation and its close friends copious data on subjects that do not directly affect its own immediate interests. The committee assignments and special interests of legislative partners must be known, and efforts must be made to bring legislators into face-to-face contact with research spokespersons, public service directors, and others who can give them useful, timely information about areas of immediate concern, both in regular and in special sessions. Revenue projections, for example, will be of great interest to legislators who are members of the finance committees.

The rigorous examination of a governor's budget, or the provision of an alternate legislative budget, will depend on sound analysis. This may be provided, by the state finance department or the legislative fiscal office, although the latter's projections may be politically weighted and, in some instances, hard to secure (for example, when legislators do not belong to the same party as house or senate leaders). The state finance office's numbers will be tied closely to the governor's budget request. Both sources must be examined, but it is sometimes helpful to have an independent assessment of the state's revenue projections. University econometric projections may also be of use here. When a campus economics department or business center is engaged in this type of forecasting, the year-to-year projective data can be of great supplementary value to legislators whose committee assignments require them to draft or amend state budgets.

Other kinds of information will also be useful to legislative allies, including studies of state agencies and of how they can expand their services through cost-efficient reorganization and planning. Analyses of demographic trends are useful in underlining voting positions on proposed state expenditures for public services. Research on economic development, especially research that suggests ways to create new jobs, can be applied to any number of issues that command the attention of a legislative session. The funding of state schools occupies a great deal of a legislature's time and attention. Efforts should be made to encourage studies and analyses of the state's education-funding mechanism. Specific recommendations may be offered on alternate revenue and taxation measures. Environmental impact studies are useful, and some lawmakers will be interested in the effects of proposed state and private development in their districts. Items of specific interest that will emerge in a given session must be looked for and addressed. As a prelude to the assembly of researched information on such subjects, one need only talk to a number of key legislators in each party to discern what kinds of legislation they plan to introduce. It is sometimes interesting to move one step beyond this process and discover what the most effective lobbyists hope to accomplish in a particular session. Given the understaffing of most state legislative drafting offices, as well as the almost complete lack of office staffing for individual legislators, these powerful lobbies often draft the major bills that will appear in a given session. The insurance industry, trial lawyers, and the medical profession will each have interests, although they will have differing stakes. Working with one of these groups and their legislative liaisons could prove very useful, if the group's position is compatible with the university's interests. A university research center may be working toward the completion of a statewide study of the insurance industry in the year before the introduction of tort legislation. The point is this: a university has much more to offer in its lobbying efforts than football tickets and sentimentality (although one will not wish to minimize even these elements in the context of the overall program). The key to a redefined legislative relationship, just as it is in our dealings with governmental units, is our willingness to generate and share needed information.

Links should also be sought with the state finance office. The governor's budget is either drafted or approved by this executive agency, and so a campus will want access to this important decisional arena for the inclusion of its line items in the initial document. One's lines have a better chance of staying in the executive budget, once drafted, than they do getting in through amendment during the deliberations of a legislative committee. Among other things, this office and its subordinate units will be concerned with efficiency and cost containment in all government operations. Some important ancillary functions (for example data-systems management), may be supervised directly by finance. A university service program, with a record of accomplishment and cost-effective assistance to the state and its agencies, may have success here. Keep in mind that this unit will also control the important contract resources known as year-end money (the unexpended resources of a given fiscal cycle). State finance, in some cases, will be able to alert a valued, friendly institution to needs and contracting opportunities in other executive agencies and indicate when specific federal money may be available.

For a modern university, the development of a coherent and supportive governmental strategy necessitates an understanding of the political milieu and an institutional commitment to forcing favorable outcomes in the decisional arena. A campus cannot hope to assign its government program to only one person or one office and still hope to meet its long-term goals. A successful effort will be sustained, orchestrated by campus leaders, and cognizant of the informational power and support mechanisms that academe can bring to bear on the improvement of state and local governmental performance. For the university willing to examine its own commitment to public service and applied research and to reconfigure its program and service levels, some splendid opportunities exist. Today's institutions should make every effort to be alert and responsive to these options, for they will continue to be with us as our states and municipalities prepare themselves for the technical, staffing, and fiscal challenges of tomorrow. The price of success in university-governmental cooperation will be sound planning and a willingness to amplify and expand campus services in a credible and convincing manner.

James T. Kenny is vice-chancellor for research and development at Auburn University, Montgomery.

The university research center of the future will function best when a highly trained core staff, adequately equipped, provides the direct services demanded by private- and public-sector clients.

Managing a Modern University Research Center

John G. Veres III

This discussion of modern research administration begins with a plea for patience. When he was asked to contribute to this volume, the author's task was to describe the "nuts and bolts" of managing a university research center. The comments that follow have been drawn from the author's personal experiences as director of Auburn University at Montgomery's Center for Business and Economic Development (CBED). This chapter addresses a number of issues that seem important. Before exploring these management issues, a discussion of the author's early conception of research centers is presented. While perhaps not terribly enlightening to those experienced in the research enterprise, such an approach serves as an entrée to several points that should prove worthy of discussion.

A Rose Is a Rose?

Each interested person probably has a slightly different conception of precisely what constitutes a university research unit. Not too many years ago, the words evoked in the author an image of persons with scholarly mien attired in white lab coats. A few of them wear beards, perhaps a few of them smoke pipes. They all work in the physical and

J. T. Kenny (ed.). *Research Administration and Technology Transfer.*
New Directions for Higher Education, no. 63. San Francisco: Jossey-Bass, Fall 1988.

life sciences; they are physicists, chemists, and researchers in areas of biomedicine. One can find them debating conceptual formulations over the lunch table. The pace is leisurely. The professorial scientists and their graduate students march to the pace of the research itself, rather than to external pressures. In the end, discoveries are made, inventions are created, and the world is made better in some tangible way.

The outside world, recognizing these men and women of science as helpmates to the human race in its evolutionary progress, gladly supports their work. The federal government makes substantial grants available. Private foundations are not far behind, eagerly funding the most esoteric of projects. Time is not a concern. Outside funding sources patiently await the discoveries that will change our world. In a sense, time ceases to have meaning for the dedicated researcher.

Such was the author's conception of a university research center. Readers experienced with such centers and related institutes may say to themselves, "What naiveté!" And they are correct. This notion was an ingenuous one, but it is worth mentioning because it is shared, at least in part, by many inside and outside the university setting, including faculty members. Perhaps faculty do not expect white lab coats, but they do expect research centers to support their need for studies leading to publication. Many feel, legitimately, that one duty of institutes is to provide access to data that can be parlayed into journal articles.

Another major misconception concerning such units is one that may profoundly affect their operation. They are seen by many, perhaps in particular by university administrators, as income producers. It is true, that indirect costs generated by center activities represent a significant source of external revenue for many universities. All too often, however, university administrators tend to view the work of such centers as ancillary to the income such work provides. Contributing to such attitudes, hundreds of reports are published annually, praising researchers for captured dollars and signed contracts. Furthermore, supervisors of unit directors are frequently evaluated in terms of the "bottom line." Unfortunately, therefore, the mandate to bring in more money at any cost is often tacitly communicated downward to research center personnel.

Walking the Tightrope

The desire for increased income, felt by administrators, and the desire for increased research opportunities, expressed by faculty members, may appear to be diametrically opposed. This situation creates a quandary for the research center director, who may be sympathetic to both views. Many times, such units simply cannot provide faculty members with the data they desire because the service-oriented projects they undertake do not generate information suitable for publication. Moreover, many center directors are active researchers themselves. As researchers, they may well be interested in studies unlikely to attract external funding. This difference in priorities can prove troublesome.

As problematic as the research center's internal tug-of-war for resources may seem, university pressures pale in comparison to those brought to light by the external environment. Deadlines imposed by contracts, federal courts, or client fiat increasingly have become a way of life for directors and their personnel. To a large extent, centers are what they do. The character of an organization is often shaped by its work. This becomes apparent when one examines the effect of the environment on staff composition. Directors generally employ people with skills to match the demands imposed by current projects. They also tend to seek out new projects that nicely match the capabilities of center staff. Eventually, then, the selection of new projects on the basis of employees' characteristics tends to result in specialization. Specialization in one or several disciplines is fostered by a unit's undertaking increasingly complex projects in a given area. The type of project performed has implications for management style, employees' rewards, and other administrative aspects of a research center's operation.

Reflection of a Changing World

The author's center does no work in the hard or life sciences; rather, most of its work deals with the social sciences: economics, sociology, and psychology. The remainder is in areas that have not always even been considered sciences: consumers' behavior, marketing, and voters' intentions. In most cases, no tangible products are produced. The vast majority of the work is best described as provision of services.

CBED may be very much like research units at other universities. More and more, the activities of CBED focus on the timely provision of services, particularly in human resource management, automation, and organizational development. Discussions with colleagues and review of the literature have convinced the author that this unit is not alone in this regard. Today's hot topics include organizational restructuring, assessment centers, corporate climate, risk management, and a host of automation-related issues; nor are these topics simply fads. Ten years ago, Johnson (1978) examined questions on corporate social responsibility, the violation of employees' civil rights in industry, and the control of information in the corporate structure. Even then, experts questioned the efficiency of the traditional hierarchical corporate structure (Demsetz, 1978). More recently, Naisbitt (1984) examined a move from the industrial society to an information society, with its attendant implications. He concluded, "The computer will smash the pyramid: we created the hierarchical, pyramidal managerial system because we needed it to keep track of people and things people did; with the computer to keep track, we can restructure our institutions horizontally" (p. 288). Today, the CBED staff has already witnessed the impact of the phenomenon Naisbitt describes. Frequently observed activities include diagnosis of organizational problems and recommendations for structural change in clients' businesses. Fully half our projects involve automation of clients' activities, to some extent. Virtually every project undertaken is accomplished with the aid of computers. Many of the most recently signed contracts involve the development of fair and job-related devices to select or promote employees. Staff members find themselves testifying in federal court on the technical soundness of personnel systems developed by CBED or others.

Once again, this author asserts that his center is much like other units. The American Association of State Colleges and Universities (AASCU) (1987) describes cooperative projects between universities and the corporate community, which this author believes will be typical of research centers' activities in the years to come. One unit noted by AASCU is Eastern Illinois University's Community Business Assistance Center, which has developed some fifty workshops and seminars offering practical approaches to human resource problems. In a similar vein, Fitchburg State College (Massachusetts) has organized the Montachusett Economics Center to provide marketing and management assistance to regional businesses, nonprofit agencies, and governmental entities. Southern Illinois University at Edwardsville provides specialized consulting, training, and technical assistance customized to individual companies' manufacturing needs through its Center for Advanced Manufacturing, Production and Technology Commercialization.

In the Northwest, Western Washington University, through its visual communications education unit, is working with a California software firm to develop training materials for a program that coordinates word processing and typesetting operations. Moorehead State University has developed a mountaintop agricultural complex and has the means of sharing its facilities with industry. The AASCU list goes on, mentioning technology, research parks, and an assortment of institutes and centers. Direct provision of services is here today and growing.

With every day that passes, CBED less and less resembles a setting of professors in white lab coats, becoming more like the private consulting firms with which CBED often competes. The trend toward service provision will endure. This basic change in the way research centers operate will make it increasingly difficult for university faculty and administrators to understand and sympathize with the problems encountered by unit directors.

The changes of the last decade have rendered obsolete many conceptions related to operations. More than a few of these conceptions have assumed the status of myth and legend, working to limit units' ability to cope with a rapidly changing environment. The sections that follow will explore a number of popular myths associated with university research centers and discuss their implications for management.

Staffing

One myth concerns staffing. It holds that centers should be staffed almost entirely by administrative personnel. In this view, a center's role is to be a facilitator, brokering university resources to meet its clients' needs. Content-specific expertise is drawn solely from faculty members, obviating the need to employ researchers in specific disciplines. In such an atmosphere, staff members become experts in such areas as proposal writing, contract negotiation, and project management. Faculty members serve as principal investigators and research associates, with the center's personnel acting as project administrators and secretaries.

Faculty members often support this model of the university research center, for a variety of logical reasons. Consulting opportunities for university faculty are enhanced in this model. Further, such a system tends to vest more control of the research process in the hands of university faculty. The role of the principal investigator is paramount in this approach, because the university research center has no in-house expertise with which to plan and execute research designs. Other university faculty serve as research associates and consultants. Their collaborative efforts are more likely to produce projects with publishable results than projects designed strictly to solve clients' particular problems. This is not to say that research and problem solving are mutually exclusive, but this author has observed a tendency among colleagues from academic departments to design projects that are more complicated than they really need to be. This tendency is particularly pronounced when the additional components are not judged burdensome to the client and can provide results that are well suited to publication.

Many university administrators also prefer the facilitative model to one that would involve the hiring of a larger core staff for a center. The reason is simple: overhead. The cost of maintaining larger core staffs in university research centers may be unattractive to central administrators. By the same token, small staffs and increased faculty participation may appear to be a more efficient use of university resources. In at least one university with which this author is familiar, central administrators have decided that faculty members working on the research center's projects should not be compensated in the same way that external consultants are. Obviously, the use of university faculty in lieu of professional staff in the centers is very attractive under this scenario.

There are difficulties associated with the facilitative approach, however. University faculty members must schedule their activities around the classes they teach. This situation can severely hinder the progress of projects because of the associates' difficulties in scheduling the many meetings that are necessary for the coordination of efforts.

A second, time-related problem is concentration of effort. Often the federal court, in which centers increasingly find themselves involved, imposes very specific deadlines for such activities as depositions, hearings, and trials. For example, during a four-month period several years ago, CBED was hired to conduct a post-hoc content-validation study of an existing multiple choice written test. Because of time constraints beyond the center's control, this author had to work virtually every day of those four months. The vast responsibilities of teaching generally prohibit faculty members from making such substantial commitments.

When both these problems are taken into consideration, the importance of establishing a core staff in the research center becomes apparent. The CBED staff includes three Ph.D.-level industrial psychologists, who serve as principal investigators and project managers; three other professionals, who function as job analysts, project specialists, and technical editors; two graduate research assistants, who are responsible for most of the quantitative analyses; and two secretaries, who serve as the "glue" that binds our operation. Without this core staff, CBED could not respond to its clients in a timely or effective manner. The facilitative model simply would not work for CBED; moreover, this author believes it would not work well in today's demanding, service-oriented environment.

Financing

Believers in a second misconception hold that research centers function best when they are entirely self-supported. Obviously, this theory is dear to the hearts of many administrators. University officials continually search for additional funding, and every dollar not earmarked to a research center can be used for some other purpose. Faculty members may also find this argument very appealing, for much the same reason. Since the activities of the typical research center are very different from those of academic departments, some faculty members question the research center's relevance to the university's mission. A familiar argument asserts that if research centers wish to engage in activities that are, in the minds of some faculty, on the periphery of the university's central concern, then such activities should pay for themselves; they should not detract from the core missions of teaching and individual research.

Any attempt at total self-support in research centers brings up a number of problems, and first among them is staff stability. It is very difficult to maintain a core staff in situations that are totally contract-dependent. Staffs tend to grow and shrink as contracts come and go, and this situation affords little continuity to operations over time. For modern-day units to respond quickly to the external environment, they must employ people who can be free to pursue new research projects when they arise. The presence of a substantial core staff also offers an opportunity to update methods and procedures on the basis of knowledge gained through previous projects. Such updating can take place during slow periods between contracts.

One corollary of this misconception states that if a research center is not totally self-supporting, then at least it should break even, and any profits from its contracts should be put back into the university's general fund. This corollary, too, poses problems; one deals with the acquisition of equipment to supplement and augment the staff's activities. Such equipment is no longer a luxury, if it ever was, and center directors, who control whatever small profit may arise from contracts and grants, can invest these funds in new equipment and expand their center's capabilities. When this author first became CBED's director, the unit possessed one so-called word processor that had a twenty-character display screen. Six microcomputers were listed on the last inventory; and another has been ordered. Despite this increase in equipment, CBED's staff members still have insufficient access to the microcomputers, which are used as word processors and data processors.

Data-processing equipment is only one example of what is needed. Competitive units must keep pace with changing technology in their areas of expertise. They cannot provide the services demanded by a modern, information-oriented society if they have no access to the many extant databases. Subscription fees and access charges to databases have become an acknowledged cost of doing business. Relying on the university's beneficence to provide these services is all too often a poor strategy. A colleague at a major university recently sought access to a database so that he could confirm the accuracy of his work for a forthcoming journal article. The first three universities he approached, including his own, replied that access was unavailable or was restricted to certain internal departments or schools. CBED was able to provide the information, but his difficulty highlights the need for centers to ensure their own access to data, either singly or in cooperation with similar units. This is impossible without funds to invest in the expansion of a center's capabilities.

Still another problem with self-support and breaking even is the limitation such practices impose on the center director's flexibility in choosing projects. University research centers often perform public-service work, rather than research based on contract. In these instances, some or all of the costs of doing the work are donated by the research unit; for example, CBED does at least one annual project for the local Chamber of Commerce. Virtually no costs are recovered from such projects, which may still be of enormous benefit to the areas served by CBED. If CBED's finances were structured so that profits on all projects were essential, CBED could not provide public services for groups that arguably form a major part of the university's constituency.

This lack of flexibility may also come into play in the selection of paying contracts. It seems that the only thing differentiating university research centers and private consulting firms is the ability of a university research director to say no to a client. This does not imply that private consultants are more likely to take on projects with troubling aspects; this author knows that is untrue. Nevertheless, it seems that one tremendous advantage of working in a partially supported research center is the lack of economic pressure associated with project selection. Managers, who often face the choice of undertaking projects whose elements run counter to their best professional judgments, understand such pressure.

Employee Relations

Another widely held misconception, which has considerable impact on the way research centers are managed, is the belief that research centers' employees are comparable to academic employees, at least in terms of salary. Professional researchers who work in centers and hold terminal degrees tend to be paid about the same as individuals who hold the same degrees and work in academic departments. This can pose several problems. The time demands of research centers may lead research professionals to perceive inequities in pay, particularly when they must work substantially more than forty hours per week. This is not to say that all faculty members work forty or fewer hours per week, but in this author's experience, the average faculty member does tend to put in fewer hours than center professionals do. Another problem stems from the researcher's mission. Researchers' jobs involve contract work. Consulting work, taken for granted by most faculty members who perform it, may constitute a conflict of interest for researchers, who must turn it down, and this obligation may create real differences in pay.

Nevertheless, perceived inequity has not posed a major motivational problem among Ph.D.-holding staff, in this author's experience. Most people who choose careers in research centers understand the rules of the game, and salaries in academic units are also generally high enough to preclude widespread dissatisfaction among Ph.D.-holding researchers whose salaries are geared to that standard. Professional employees holding bachelor's or master's degrees, however, may well be substantially underpaid and acutely aware of this fact. In the case of those holding master's degrees, the closest academic counterpart is the instructor. Many academic instructors are essentially temporary employees, and their salaries tend to reflect their status. Moreover, pay for center staff holding only bachelor's degrees can be a tremendous problem for center directors, because there are virtually no individuals holding only bachelor's degrees who work in typical academic units, and so there is no standard for compensating center employees who hold bachelor's degrees. In many research centers, master's- and bachelor's-level employees also form the backbone of the work force. They bear the brunt of the long working hours necessitated by stringent deadlines. Thus, center directors may not be able to offer meaningful salary-based rewards to staff below the Ph.D. level.

This situation is aggravated by the efforts of many universities to lower the overhead costs of total operations. Quite justifiably, administrators have sought to funnel available salary funds to faculty members, who are so essential to universities' central mission. At some institutions, this practice has resulted in lower salaries for administrators. Research professionals may find themselves lumped in with these others, because many universities dichotomize the professional work force into faculty and administrators. This trend further limits a director's use of salary as a management tool.

Faced with such salary-related difficulties, many center directors have focused on other methods to motivate their employees. Porter and Lawler (1968) differentiate between extrinsic rewards and intrinsic rewards, a distinction that directors should understand and value. Extrinsic rewards are given by the organization. Salary, fringe benefits, status, and working conditions are examples. Intrinsic rewards are self-administered. They include self-validation for a job well done, satisfaction derived from the work itself, responsibility, and personal growth. Those who want to animate research must use extrinsic rewards, other than salary, as devices for motivating employees. One such device is to ensure that each professional employee has access to the tools and equipment necessary to accomplish tasks. Computers, dictaphones, and other items can greatly assist professional employees in the performance of their duties. Even when money is not available for salary increases, center budgets may well accommodate the equipment purchases that can dramatically improve morale. Microcomputers on staff members' desks constitute a perquisite of considerable value. Such perquisites are extrinsic rewards, not subject to the vagaries of salary constraints.

Another kind of perquisite is the working environment itself. Carpets on the floors, flat latex paint instead of institutional enamel on the walls, and attractive furnishings can materially affect the way employees view their work. Improving the appearance of the work environment can also enhance a center's image with clients who are more at home in the private sector than in government institutions. Pleasant working surroundings are one extrinsic reward available to the center directors whose ability to increase salaries may be limited. Access to university vehicles is another such reward. Universities often pay mileage compensation to employees for the use of private vehicles, but many employees prefer to avoid the wear and tear on their own automobiles for intracity and out-of-town travel.

Of course, directors always have at their disposal the intrinsic rewards identified by Porter and Lawler (1968) and others. Such devices may be the most valuable management tools. Hackman and Oldham (1975) list five core job dimensions, which lead to the intrinsic rewards noted by Porter and Lawler and ultimately to high-internal work motivation, job satisfaction, and high-quality work performance. These are task significance, task identity, skill variety, autonomy, and feedback. Significance, the first dimension, is the quality of the work almost inherent in a research center. Centers solve real problems for their clients, and employees tend to view their tasks as important contributions. Skill variety, task identity, autonomy, and feedback are four qualities of job duties that may vary considerably, and which are very much under the control of center directors.

Some of Hackman and Oldham's observations bear repeating, although they may seem self-evident. Center employees work best when they are assigned tasks that allow them to use a variety of different abilities, skills, and talents. They work best when they are allowed to do the whole job, rather than only a portion of it. Their performance can be improved if they have considerable freedom in approaching their work, discretion in scheduling their activities, and choice of means for accomplishing particular jobs.

On the Folly of Following Free Advice

A wise man once said that free advice is worth every penny you pay for it. One could briefly summarize this chapter's advice, for what it is worth, as follows:

1. University research centers function best when they are composed of core staff who possess education and experience in the area(s) in which the center plans to offer services.

2. Centers should be financed with a combination of hard monies, to ensure stability in core staff, and profits generated by the center's activities. Profits should be retained by the center to build capacity, rather than being passed on to the university administration. The latter should content itself with the indirect revenue generated by centers' activities.

3. To maintain high morale and quality of work performance, center directors should earmark a significant portion of revenues for the acquisition of equipment, which increases the efficiency of the staff, and of furnishings, which enhance the working environment. Furnishings should be comfortable enough to represent a viable extrinsic reward for employees.

Having freely offered this advice, this author feels a certain responsibility to predict, strictly on the basis of his own experiences, some of the outcomes that can be expected, other than those previously mentioned. The first of these is jealousy. The plethora of equipment and work-related tools necessary to increase efficiency and boost morale may invite comparisons. Some universities are unable to provide their faculty members with the equipment and resources that research directors should view as necessary. Access to data, enhanced computing capabilities, and other clearly job-related perquisites of working in research centers can engender strong emotional reactions from some faculty in such institutions.

Faculty's dissatisfaction may also be aggravated by differences in working environments. Improvements in the center's environment serve both to motivate employees and to make clients feel more at home in the center, such improvements may be perceived by faculty in a very different light. One occasionally hears stories of "luxurious accommodations" and "executive suites." Particularly in an environment perceived to be poor in resources, researchers may come to be viewed as landed gentry living off the hard labor of the working class.

Of course, there are things that center directors can do to defuse criticism. Equipment can be purchased for university faculty with the center's budget surpluses. Proposals and journal articles can be typed for faculty, and faculty can also be sent to conferences. Publicizing the daily aspects of the center's work in university publications may ameliorate the attitudes of noncenter employees toward the center. This author has tried all these things. In some ways, these techniques have reduced negative perceptions of the center, but substantial alteration in attitudes takes time. Research units can peacefully coexist with academic departments, but patience is required of all concerned.

The future presents a number of challenges for campus research units. We can make the transition to service-oriented operations. Many centers are already actively pursuing service-based enterprises (American Association of State Colleges and Universities, 1987). Others will follow. Can we make our colleagues understand the necessity to diverge further from traditional academic themes? One hopes so. Research centers and academic units can only benefit from cooperation, despite our normative and operational differences. Indeed, our differences can and should lead to a broader conceptualization of what constitutes the modern university.

References

American Association of State Colleges and Universities. *Exploring Common Ground: A Report on Business/Academic Partnerships.* Boston: American Association of State Colleges and Universities, 1987.
Demsetz, H. "Are Large Corporations Inefficient?" In M. B. Johnson (ed.), *The Attack on Corporate America.* New York: McGraw-Hill, 1978.

72

Hackman, J., and Oldham, G. "Development of the Job Diagnostic Survey." *Journal of Applied Psychology*, 1975, *60*, 159–170.

Johnson, M. B. (ed.). *The Attack on Corporate America*. New York: McGraw-Hill, 1978.

Naisbitt, J. *Megatrends: Ten New Directions Transforming Our Lives*. New York: Warner Books, 1984.

Porter, L. W., and Lawler, E. E., III. *Managerial Attitudes and Performance*. Homewood, Ill.: Irwin, 1968.

John G. Veres III is director of the Center for Business and Economic Development at Auburn University, Montgomery.

University administrators may not recognize their own research offices within ten years unless they begin to plan now for the changes that new cooperative research programs will bring.

The Future of University Research Administration

Mark Elder

The one constant in the short history of modern research administration American universities is change. Perhaps this can best be illustrated by the fact that the word *grant* meant *gift* as late as the early 1960s but now essentially means *contract.* Accordingly, the administration of research grants has changed from the principal investigator's spending funds essentially as he or she chose to a complex system of technical and financial project oversight, requiring accountability at all stages.

Since most university research was and is funded by federal government agencies, accountability for the expenditure of public funds should have been expected by university research administrators. That it was not—and is still resisted by some—clearly shows that the focus of research administration has been and still is on the past; that is, how can we "deregulate" research and return to the good old days of minimum oversight?

Clearly, our focus, like that of our scientists, should be on the future, for there are major changes coming, not only in the way research will be funded but in the way research programs will be administered. This chapter postulates some of those changes, in the hope that we can alter our focus and work with government and industry to make research administration an actor, instead of a reactor, in those changes.

J. T. Kenny (ed.). *Research Administration and Technology Transfer.*
New Directions for Higher Education, no. 63. San Francisco: Jossey-Bass, Fall 1988.

Major Determinants of the Future

Barring a major military conflict, funding for research at American universities will be most affected by the U.S. economy. In the near future, increased attempts to reduce the national debt, balance the budget, and lower trade deficits will combine to shrink the funds available for long-term basic research. These same factors may also make more funds available for short-term applied research in some areas, especially if such research promises new products and processes that may help stimulate the economy and reduce the trade deficit. New arms-reduction treaties and a new president may result in some shifting of funds away from military research, but if this happens, those funds will most probably be used to reduce the deficit. The overall result will be, at best, a reduction in the total funds available (in terms of constant dollars) for university research during the next several years.

Unlike previous reductions in research funding, this reduction will affect not only individual projects but also the way research itself is administered at universities. Until now, academic research administration has been virtually isolated from world market conditions; funding increases, reductions, and shifts have not significantly altered what we do or the way we do it. The current situation is different. For several years now, universities have argued that their research is a major economic development tool and should be funded as such. Federal and state agencies have now accepted that argument, and our future research productivity will be judged in large measure by how much technology we generate, how we transfer that technology, how many new companies are started from our technology, and how many new products are derived from university research—in short, what we add to the economic competitiveness of our states and our nation. Likewise, major research funding will increasingly be awarded on the basis of programs' potential contributions to American competitiveness (real and perceived) in the world market.

Obviously, the demand for competitiveness will not override all other needs. That demand will be tempered by some very real political and social constraints in the United States. For example, the U.S. population is aging, and persons over fifty consistently vote in the largest numbers. Congress is well aware of this potent combination and thus will continue to fund this group's priorities, which include health-related research. Similarly, strong lobbying by other special-interest groups will result in a shifting of some funds to those politically attractive areas. Major new research programs funded by the federal government, however, are most likely to involve basic and applied research that offers good potential for new, commercially feasible technology.

Models for New Research Programs

What sorts of programs are these likely to be? We already have a number of potential models from Japan and Europe. In 1981, Japan announced a twelve-year, $500 million project that created a nonprofit corporation to link eight companies together for a "fifty-generation computer project." Since that time, Japan has formed additional projects in biotechnology and microelectronics. In 1983, the United Kingdom launched a five-year, $300 million program for advanced information technology involving large-scale integrated circuits, computer programming, better person-machine interfaces, and systems that can apply knowledge through inference. The British government also started two companies to commercialize biotechnology. Also in 1983, the Federal Republic of Germany began its own microelectronics research program, funded at $1.1 billion.

Attractive as Japan's model may seem, the United States is unlikely to emulate Japan's programs. American society has historically divided the roles of government, industry, and universities much more than European and Asian countries do. With histories of state-owned or state-controlled industries at various times, those countries have few qualms about selecting small groups of companies and funneling large amounts of government money into them to achieve stated goals. Americans differ significantly because of our history of free markets and open competition, not just with other countries but also among ourselves. Our philosophy is that everyone should have an equal opportunity to compete, and that the open market will produce the best products at the best prices. We have even codified these beliefs in our antitrust laws. Although previous special projects, including the Manhattan Project and the Apollo space program, have modified our basic beliefs somewhat, to adopt the Japanese model in its entirety would require us to ignore major aspects of our history, beliefs, and laws to an unacceptable degree.

How, then, will the United States respond to these challenges? The United Kingdom probably provides a much better model for us than Japan does in this regard. The United Kingdom has outstanding research universities, which it employs to the fullest. The Federal Republic of Germany also is developing its research universities as partners for government-funded industry-university research programs. At present, American research universities are stronger, more numerous, and better equipped, on the whole, than those of any other country, although plant and equipment obsolescence is rapidly approaching a critical stage. Together, these universities constitute America's "ace in the hole."

This view is supported by the Microelectronics Computer Corporation (MCC) and Sematech, two programs in Austin, Texas. MCC, a national consortium designed to create new computer architecture and software, is funded solely by industrial participants and thus is not a government-industry venture. Sematech, which will develop and test advanced manufacturing processes, materials, and equipment for the U.S. semiconductor industry, is supported by the federal government, industry, the state of Texas, and the University of Texas at Austin. Sematech's operating budget will be provided by both government and industry. The state of Texas is providing the primary facility for Sematech, and much of the Sematech research will be done at the University of Texas at Austin, in research laboratories. Thus, Sematech will benefit from working closely with industrial firms, a sister consortium, and a major research university.

These consortia also illustrate changing U.S. attitudes toward such ventures under the pressure of global competition. When MCC was formed in 1983, questions were raised concerning its legality, and special legal opinions were necessary to avoid antitrust lawsuits. To date, no one has challenged Sematech (founded in 1988), despite its much broader implications for free-market advocates.

New Federal Government Programs

In the future, the federal government will most likely use its funds, influence, and regulatory power to foster greatly increased cooperation among universities, industry, and government laboratories. This will take several forms, ranging from simple redirection of funding toward cooperative projects to creation of specific government-industry-university consortia to accomplish targeted objectives. The federal and state governments alike will establish such programs, but state programs will focus on more narrow regional concerns that promise to broaden tax bases and quickly reduce unemployment.

Within the next few years, the federal government should introduce major new research programs in scientific fields from which substantial economic and health benefits can be expected (for example, in biotechnology, molecular genetics, materials sciences, and neurobiology). Breakthroughs in individual areas within fields will also generate new research programs. A good example of such an area is superconductivity, where one initial breakthrough had led to nothing less than startling advances in the state of the art. This area alone offers vast increases in energy efficiency, along with such potential developments as competitively priced electric automobiles, high-speed trains levitated by magnetic fields, and better research tools (for example, the superconducting supercollider).

American research programs will concentrate on cooperative ventures, within a framework of free-market competition, to promote scientific advances. This will be accomplished by several methods intended to lure or force government agencies, industrial firms, and research universities into new forms of cooperation. These methods will include the funding of large programs and the use of regulations against those that refuse to share new knowledge and new technology.

Such programs cannot be established solely by funding and fiat; universities and industries must see something worthwhile in each program for themselves. Further, the benefits must be sufficient to overcome decades-long suspicion and distrust between the parties. This means that university scientists and engineers must be willing to take some outside direction on their project schedules, and must be willing to allow outsiders into their laboratories as full participants in the work. The latter point is absolutely essential, both for quick, full disclosure to other parties and for instantaneous transfer of technology to industry. For industry, it means giving up significant amounts of proprietary information to outsiders, allowing company scientists to work outside company facilities (where their supervisors may or may not know what they are doing), and making significant investments of company funds and personnel in projects that do not promise new, company-owned products for the market by specified dates. This constitutes a major gamble for most company executives, who must find a way to convince both their cost accountants and their stockholders that they are truly investing in the future.

Effects on Universities

The effects of the U.S. economic situation and of increased competition on university research are already evident in several areas. First, the nondefense funding of university research has plateaued or decreased (in terms of real dollars adjusted for inflation) during the last several years. Second, overall public support for higher education has decreased in many states, because of stagnant economies and additional service demands. Third, student enrollment is up at virtually all major universities, further stretching state and private university resources. Fourth, demands for increased accountability have caused universities to devote more resources to complying with those demands. Fifth, major industries and small businesses are now competing for funds that once were the sole province of research universities.

This combination of factors has produced a number of results. For example, the increased competition for funds has produced more "big science" projects and transdisciplinary university laboratories that concentrate on applied research. These programs bring their own forms of stress to researchers, creating an atmosphere in which they may feel that they no longer can afford to fail. In some cases, one result has been the trimming of data to fit hypotheses, and other forms of scientific misconduct. Another result is direct lobbying of Congress by individual universities for funds to build facilities, update equipment, and obtain major projects. Although such lobbying is decried by the Association of American Universities and other major university-based research groups, more and more universities are now quietly conducting their own lobbying campaigns to remain competitive. Since there are few if any sanctions that can be applied to individual members of most university associations, lobbying efforts by universities will probably increase during the next few years. These are only a few of the current results. Future research programs conducted cooperatively by government and industry will produce other, more dramatic changes, such as more government regulations, new forms of cooperation, increased accountability, and substantially different management policies and procedures.

Increased Government Regulation

University research administration will be changed significantly by targeted cooperative research programs, but not necessarily in the ways that many administrators expect. We will continue to manage individual, government-sponsored, university-controlled projects, the administration of which varies primarily only by amount of funding. We will also administer interinstitutional, university-industry, and university-government projects. Changes in these areas will be evolutionary, although they will occur much faster than previous changes.

With increased government funding, there has been greater demand for public accountability. Until the late 1970s, this accountability primarily took the form of postaward fiscal regulations that restricted expenditures, specified detailed financial reports, and required regular audits. Since that time, however, public concerns have produced new regulations, with both preaward and postaward responsibilities for human subjects, animal research, and biohazards.

The restrictions on research involving human subjects peaked early in this decade and have since been relaxed to some degree. For example, at one point the Office for the Protection from Research Risks of the U.S. Department of Health and Human Services required universities' institutional review boards to review and approve both survey research and purely observational research. Those requirements have now been relaxed, to the point where most such research is now exempt or can be approved through expedited review procedures.

This is not true of animal research, however. Public concern, fed by real and perceived abuse of animals, has shifted toward increased protection of vertebrates, principally warm-blooded mammals. The results are new and increasingly stringent policies from the U.S. Department of Agriculture and the Public Health Service of the U.S. Department of Health and Human Services. Primate studies have been the focus of concern, but in the future, better-funded and more sophisticated public information campaigns by moderate and radical antivivisectionist groups may result in the expansion of these regulations to cover additional species, such as rats and mice. New regulations may also require the substitution of nonanimal models, such as cell cultures and computer models, whenever possible.

In addition to human and animal research regulations, Congress has enacted the Export Administration Act of 1979 and the Export Administration Amendments Act of 1985, as well as the Arms Export Control Act. These acts regulate the types of technology that can be sold or otherwise transferred to unfriendly nations, even through companies in friendly nations. A prime example of the reach of these acts involves the sale of computers by a division of the Toshiba Company of Japan to the Soviet Union. These computers will allow the Soviets to design submarine propellers that are less noisy than current Soviet designs, thus making detection of Soviet submarines much more difficult for American antisubmarine devices. The U.S. Department of Defense was so disturbed by this technology transfer that several congressmen suggested banning some or all Toshiba products from U.S. markets for several years.

Unless a new period of détente between America and the Soviet Bloc countries develops from current arms negotiations and treaties, we can expect the Export Administration Act's restrictions to be broadened. Further, similar legislation could easily be used to prohibit the transfer of not only defense-related technology but also economic technology. Should this occur, the impacts on university research may involve restrictions on the participation of foreign students in university research or the licensing of technology transferred to foreign corporations.

Controls on direct technology transfers have already been extended to include some indirect technology transfers. These have come primarily in the form of barring certain foreign students from participating in certain defense-related projects that could give them information about sensitive (but unclassified) technology, which they could take back to their home countries. These controls may be expanded. The federal government (and some state governments) may also begin to limit fellowship support for students from countries deemed both "unfriendly" and "friendly."

As Congress, state legislatures, and the public become more knowledgeable about the methods used by some foreign countries to get their students educated in the United States, public pressure may grow to limit support for such students. For example, China has a standing policy of providing only one year of support for its students attending U.S. universities. After that initial year, the students must find sufficient funding from scholarships, fellowships, or research assistantships to support themselves; otherwise, they must return to China. This policy has worked very well for China, because the students sent here major primarily in the sciences and engineering. With a dearth of American students in those areas, university research faculty must accept foreign assistants or forgo many worthwhile projects. If the use of foreign students on research projects is limited by law, regulation, or policy, university research will suffer accordingly. Thus, without a major increase in American students taking scientific and engineering courses, this area will pose a knotty problem for Congress, research-support agencies, and universities.

Near-Term Impacts on Research Administration Offices

The near-term impacts of increased government regulation and accountability on universities' research administration offices can be summarized in one word: computerization. Most research development (pre-award) offices already use computers to some extent; grants-accounting (postaward) offices use them extensively. The reason for this imbalance is obvious: Digital computers were originally designed as accounting (and then as scientific number-crunching) tools, with word processing, database manipulation, and other uses as secondary functions that have developed more slowly. The future, however, will see a much broader use of computers, especially microcomputers, in both types of offices.

Such uses are already in the experimental stage. For example, the National Science Foundation (NSF) is currently funding and participating in a preaward experiment with the University of Michigan and Carnegie-Mellon University on the use of computers to transmit proposals electronically. The purpose of this program is to develop a system capable of linking the different computers and software systems used by universities with those used by NSF, to make such transmittals possible and efficient. The program is focused on compound documents, which include not only narrative information but also graphs, figures, pictures, and other such materials. Likewise, a postaward experiment in Florida involves five federal agencies (the Department of Agriculture, the Department of Energy, the National Institutes of Health, the National Science Foundation, and the Office of Naval Research), the nine higher educational institutions in the state of Florida's system, and the University of Miami. This project is attempting to develop a common set of terms and conditions for grants and contracts awarded to universities. The purpose is to reduce both government requirements for approvals by agencies and overall grant complexity.

Once common terms and conditions have been developed, the next step will be for universities to code those regulations into computer checklist programs, which will automatically compare planned expenditures with applicable regulations. This will require the coding of budgets for awards (with allowable expenditure fields and amounts) in such a manner as to make them comparable by computer. Such programs will provide the fiscal accountability desired by federal agencies and universities while freeing postaward personnel to work on more complex problems. The result will not only provide increased accountability but also reduce the number of postaward personnel required to administer awards and the university's indirect costs.

Universities are also developing other computer programs, both for in-house and external functions. For example, most research universities now have database programs for proposals, awards, and faculty's research interests. Some universities also have databases of their research facilities and equipment. At present, most of these programs run only simple searches and produce only basic reports. The faculty research-interest database is used primarily to locate individuals to respond to requests for proposals.

In the future, the more successful research offices will perform analyses of faculty interests to determine patterns, match university resources with regional and national programs, and develop research teams to propose major multi-institutional projects. These offices will also provide forecasts of potential funding, by field, and of the probability of funding in specific research areas. These forecasts will be used by university administrators to plan equipment purchases, hire faculty and support personnel, and construct facilities.

In the shorter term, computers will be used to provide faculty with up-to-date information on sponsored project opportunities. For example, on regularly scheduled days, a research office worker will dial a national database or one of the new, government-maintained databases. New project opportunities, deadlines, response requirements, and other selected information can be obtained and matched with information in the faculty research-interest database. When that is done, the computer will automatically send notices to identified faculty via local electronic mail networks. Such systems are fully feasible right now; the only requirement still not commercially available is translator software programs to match keywords and numeric fields in national programs with the codes used in local programs. When these systems are fully operational and widely used, national and local newsletters alike will become virtual relics for all but the most hardened print lovers. The costs of the new system will be largely covered by subscription budgets for printed materials.

The system will also allow searches by external users seeking matches for joint programs, consultants, and other needs. These searches will initially concentrate on faculty interest and expertise and then will widen to include facilities and specialized equipment. Further, they will be two-way systems, and individual institutions will be able to dial state and national databases of government and private-sector resources.

According to the policy of the individual institution, searches may be conducted through the research office or directly on line by the external user. In either case, the databases will allow quicker and better matching of university, industry, and government resources than current methods do. When these systems are built, individual faculty must have the opportunity to lock out their information to all but designated users.

Other computer programs will tie all these databases together, providing automatic checks on proposals as well as on awards. These areas will include approvals for human-subject and animal research, recombinant DNA guidelines, biohazard considerations, and current as well as new regulations. When a proposal is electronically sent through the signature route, checklist information will be provided at each step, to ensure that all necessary requirements of the institution and the potential sponsor have been met. This will not only solve a number of existing problems but will also speed proposal routing, as well as transmission to the agencies. For example, members of approval committees, such as institutional review boards, will be able to set aside regular times of as little as thirty minutes each week to review protocols deposited in their databases. The chair will send a copy to each member's computer when the protocol is received for advance review. At the designated times, the board members will use computer conferencing to go through the agenda, without having to meet physically. Members traveling in other parts of the nation, or even in foreign countries, can also participate in these reviews, through telephone links and portable microcomputer.

Proposal and award negotiations will be conducted via transmission of data between computers. Compatible, interactive budget-calculation programs will allow principal investigators and agency program officers to play "what if" with potential award budgets, obtaining (and storing for comparison) the total cost of various project configurations via instant recalculation of personnel, equipment, supplies, and other costs. Voice links through the computers will retain the necessary personal discussions, with results transferred by computer to the university research office for review and official approval.

When the program officer receives university approval on the revised budget, he or she will transmit the award information to the appropriate grant officer. This officer will add any special award conditions to his or her transmission, which the research office negotiator can read and respond to before official transmission of the award. When the award is officially received, the research office will route it electronically to the postaward office and extract the title, the budget, and other pertinent information for deposit in database and transmission to other relevant university offices. Many other such changes are possible, but these examples should suffice for a look at the needs of research administration offices in the near future.

Major Changes in Research Administration

All these changes may seem major to many research administrators who are now mired in paperwork, but those changes will be evolutionary, not revolutionary. New databases, electronic transmission of proposals and awards, and even computer checking of proposals and awards will not change the underlying management principles we now follow. Those principles will be changed, however, by the new research programs that must be developed if the United States is to remain competitive in scientific and technological fields. In short, we cannot facilitate and manage development of the sixth, seventh, and eighth generation of computers (or biotechnology, or any other high-technology field) with second-generation management techniques.

In modern times, major changes in management techniques have occurred only as a result of the industrial and communications revolutions. The industrial revolution developed vertically structured but still decentralized management techniques for factories and businesses. The communications revolution, combined with the use of computers, has allowed top managers to centralize major decision making in small groups of executives. Even the matrix-management schemes developed by the aerospace industry during the 1960s did not constitute a major change in management techniques.

University administration has not changed even that much. Major academic decisions, although officially made by boards of regents, still require the support of the faculty if they are to succeed fully or quickly. Thus, because they are generally more decentralized, universities may require less change than industrial firms and government agencies to meet the new management demands of future research programs. Those programs will require decentralized decision making, with many major decisions made at the project level.

Major new "big science" programs will be cooperative ventures of government, industry, and universities. The only individuals who can determine what those programs should be are the scientists and engineers who are now working in the trenches. Managers and executives who rose from the scientific ranks will continue to play major roles, but those roles will shift even more toward communicating and selling the ideas hatched in the laboratory. The primary functions of administrators will be to give form to programs, help set goals, find funding, and facilitate the work performance of scientists and engineers. Decisions on program directions, amount of funding required, technology development, technology transfer, and other major functions must be made by scientists and engineers. To do this, they must work together in ways that have not been possible in the past, and university research administrators must help create methods for such interactions.

The first thing university administrators must do is prepare to give up a portion of our control over research projects. Our counterparts in industry and government must do the same, so that what emerges is a form of mega-administration, in which we communicate our plans, as well as our individual decisions, to others in the loop. Again, this will be done electronically, under a basic set of ground rules set by the federal government. The overriding concern in this system must be facilitation of the work, rather than adherence to a set of codified regulations. We will have more freedom to act under the general guidelines for these programs, but that freedom will bring greatly increased responsibility for communication of our actions to others. These actions will be checked, again by computers, as they are taken, so that actions that fall outside the set limits are identified and can be reviewed and approved by the mega-administration immediately.

Financial resources will be likewise pooled, with the individual participants drawing from the primary sources on the basis of immediate need. These resources may be supplied by government, industry, or universities, which will require considerable changes in university purchasing, payroll, and other systems. For example, a university participating in a major program might have scientists paid from three sources, technicians paid from two different sources, and students paid from yet another source. Likewise, equipment might be purchased through an industrial firm not directly involved in the university's immediate work, or the equipment may simply be supplied by that firm, on or off campus, with use-time charged back to the university. Obviously, electronic fund transfers through major banks and government channels will be a key element of this system. One government agency may also be designated to act as a central bank for such projects, pooling all commitments and paying all charges incurred by the program.

Other major programs will involve international cooperation, with problems of languages, currencies, and other factors that must be addressed. These programs, like the American programs, will be driven by decisions made by the scientists, and research administrators must be prepared to facilitate the necessary administrative steps to keep the programs on track. Language-translation software may be perfected by that time; if not, our successors will need to be conversant in several languages.

More immediately, we must begin reformulating our policies on patents, copyrights, and other means of protecting and transferring technology. In the case of new processes and devices developed through cooperative ventures among university, industry, and government scientists and engineers, a major question to be resolved is who owns them. Just reaching an equitable agreement, on who contributed what and how much to the invention, will be a major accomplishment. To determine which parties have rights, what those rights are, and who will receive monetary and other benefits from the commercialization of the new technology will require the wisdom of more than one Solomon. University administrators must participate fully in the formation of new laws, regulations, and policies in these areas to ensure that our scientists receive due personal credit and monetary rewards. We cannot afford to be obstructionist while doing so, however, or those decisions will be taken out of our hands.

Likewise, we must also participate fully in the transfer of technology from our institutions to industry and government. This should be much easier in actual performance. If scientists representing all three parties work together, the transfer will occur naturally, but documenting such transfers (as a basis of later decisions on patents, copyrights, and so on) will become even more difficult. New systems will be needed, because keeping written laboratory notebooks and using other current techniques will be too cumbersome, as well as insufficient, to provide the necessary data for such decisions.

Summary

New and major cooperative research programs involving personnel from universities, industry, and government will be developed to help our nation meet competitive challenges in the world market. These programs will require decentralized decision making by the scientists involved and thus will require major changes in the policies and procedures currently used by university research administrators.

In the short term, computerization of many current preaward and postaward functions will significantly speed up tasks, provide additional accountability, and reduce the number of personnel required to administer university research. University administrators must begin preparing now for both short- and long-term changes if they are to be actors in, instead of reactors to, these processes.

Mark Elder is assistant vice-president for research and director of sponsored projects at the University of North Texas. He served as president of the National Council of University Research Administrators in 1983.

Index

A

Aburdene, P., 47, 48

Academic communities: and Batelle Development Corporation, 31; and focused agents, 31; and incentives and monetary distribution policies, 33; and national licensing agents, 30; operating models of, 29-30; professional associations and societies in, 29; and Research Corporation, 31; successful research funding in, 32-33; trends in, 29-34; and WARF, 31; and WRF, 31-32

Agricultural Extension Act, 6

Agriculture Department, 78-79; and research grants, 80-81

Alabama: and corporate/university partnerships, 43-45; and government/university partnerships, 49-59; State Employees Association of, 51; State Personnel Department of, 51; and vocational-technical education, 27

American Association of State Colleges and Universities (AASCU), 40, 48, 64, 71

American product, life cycle of, 10-11

"America's Leanest and Meanest," 38, 40, 48

Apollo space program, 75

Arkansas, and vocational-technical education, 27

Arms Export Control Act, 79

Association of American Universities, 78

Atomic Energy Commission, 7

Auburn University, Montgomery (AUM): Center for Business and Economic Development (CBED) at, 43-45, 61-72; Center for Government and Public Policy at, 50-51; and certified public manager (CPM) program, 51; and Monsanto Agricultural Company, 43-45; and state and local government, 49-59

B

Batelle Development Corporation, 30-31

Batelle Memorial Institute, 31

Ben Franklin Partnership. *See* Pennsylvania

Biological Materials Distribution Center, 31

Birch, D., 28, 35

Boston, and entrepreneurial activity, 28

Brazil, and higher education, 8

Britain. *See* United Kingdom

Brown v. Board of Education, 13

C

Carnegie-Mellon University, 80

CBED. *See* Center for Business and Economic Development

Center for Business and Economic Development (CBED): at Auburn University, Montgomery, 43-45; quality-of-worklife (QWL) program at, 43-45

Center for Government and Public Affairs, at AUM, 50-51

Center for Process Analytical Chemistry (CPAC), 26-27

Certified public manager (CPM) program, at AUM, 51

Chase Econometrics Services, 53

China, 79-80; and economic growth, 15

Cold War, 5, 13, 20, 74

Commerce Department, 11, 14-15

Congress, 79-80; eightieth, 7; and university research funding, 74, 78-80

Cornell University, and corporations, 42

Corporate Cultures, 47

Corporate readaptation: concept of, 37-38; future of, 47-48; in higher education, 37-48; higher education's role in, 40-43; and partner-

Corporate readaptation *(continued)* ships with universities, 46–47; in the past decade, 38–40; potential problems of, 45–46; university-corporate partnership in, 43–45
"Corporations on Campus," 26, 35
Cottrell, F. G., 31
Crown Cork and Seal, 40

D

Databases, 80–85
Defense Department, 7, 12, 14–15, 79
Demsetz, H., 63–64, 71
Department of Energy, and research grants, 80–81
Douglas, W. L., 6
Duke University, 28; research staff at, 42
Dyer, D., 38–39, 48

E

Eastern Illinois University, Community Business Assistance Center at, 64
"The Economy of the 1990s: Where to Live—And Prosper," 29, 35
Europe: industry after World War II in, 26; and new research projects, 75; and research-oriented institutions, 6; and world arena of power, 9
Experimental Stations Act. *See* Hatch Act
Export Administration Act of 1979, 79
Export Administration Act of 1985, 79

F

Federal Circuit Court of Appeals, 25
Federal government, funding for research, 76–85
Federal policies: and federal court decisions, 25; important new, 23–25; responses to, 24–25; and revised patent laws, 23–24; university responses to, 24–25
Federal Republic of Germany. *See* Germany

Federal Reserve Board, 37
Fitchburg State College, Montachusett Economics Center at, 64
Florida: research grants in, 80–81; and vocational-technical education, 27
Fortune 500 companies, 28
Fred Hutchinson Cancer Research Center, 28
Future of research administration, 73–85; and effects on universities, 77–78; increased government regulation and, 78–80; major changes in, 83–85; major determinants of, 74; models for, 75–76; near-term impacts on, 80–83; new federal government programs for, 76–77; in summary, 85

G

General Accounting Office (GAO), 23
General Motors, 39–40
General Motors v. *Devex*, 25, 35
Georgia-Pacific, 39
Germany: and high-technology products, 10; and new research projects, 75; and research-oriented institutions, 6
Gorbachev, M., 9–10, 15
"Growing Places," 27, 35

H

Hackman, J., 70, 72
Hammond, G. S., 22, 35
Hatch Act, 6
Health and Human Services Department, 78–79

I

India: and higher education, 8; science and engineering graduates in, 8
Indiana, and high technology, 27
Industrial Revolution, 6, 16
Iowa, and high technology, 27

J

Japan: commercialization of technology in, 22; and economic versus

military power, 12; as future consumer, 14; and high-technology products, 10; and higher education, 8; industry after World War II in, 26; and new research projects, 75; and product commercialization, 15; Toshiba Company of, 79; and world arena of power, 9
Johns Hopkins University and Hospital, 6
Johnson, L. B., 5
Johnson, M. B., 63, 72
Jonas, N., 26, 35
Justice Department, 27

K

Kellogg Foundation, 14
Kennedy, A., 47
Kenny, J. T., 38, 48
Korea, and smokestack industries, 10

L

Lawler, E. E., III, 69, 70, 72
Lawrence, P. R., 38-39, 48
Liability, 34
Licensing Executives Society, 29
Little Marion Laboratories, Inc., 40
Lobbying, 55-59, 78

M

Manhattan Project, 7, 75
Massachusetts Institute of Technology (MIT), 26; and corporations, 42; technology transfer offices at, 30
Mergers and acquisitions, of large American firms, 38-40
Mexico, science and engineering graduates in, 9
Michigan, and venture capital, 27-28
Microelectronics Computer Corporation (MCC), 75-76
Microelectronics Innovation and Computer Research Center, 42
Mikuni, A., 14, 20
Monsanto Agricultural Company: and AUM, 43-45; and Washington University, St. Louis, 26, 42-45
Monsanto-Washington University, St. Louis program, 26; and biotechnical research, 42

Moorehead State University, agricultural complex at, 64
Morrill Land-Grant Act, 6

N

Naisbitt, J., 47, 48, 72
National Aeronautics and Space Administration (NASA), 17
National Association of College and University Attorneys (NACUA), 29
National Certified Public Manager Consortium, 51
National Council of University Research Administrators, 29
National Institutes of Health, and research grants, 80-81
National Science Foundation (NSF), 22, 26-27, 80; and research grants, 80-81
Newtonian mechanics, and atomic physics, 16-17
Nixon administration, and NASA budget, 17
North Carolina: Research Triangle Park area of, 28, 42; and vocational-technical education, 27
North Carolina State University, 28
North Carolina University, research staff at, 42

O

Oak Ridge National Laboratory, 7
Office of Naval Research, and research grants, 80-81
Office for the Protection from Research Risks, 78
Office of Technology Transfer (OTT), 24
Ohio, Thomas Edison Program of, 27
Oldham, G., 70, 72
Omnibus Trade and Competitiveness Act, 11
Overman Act, 7

P

Pacific Basin, and world arena of power, 9
Patent and Trademark Amendments of 1980, 23

Pennsylvania: Ben Franklin Partnership of, 27–28; Pennsylvania Technical Assistance Program of, 27–28
Porter, L. W., 69, 70, 72
Public Health Service, 78–79
Public Law 96-517. See Patent and Trademark Amendments of 1980

Q

Quality-of-worklife (QWL) program, at CBED, 43–45

R

Ramo, S., 22, 35
"The Renewal Factor," 40, 48
Research Corporation (RC), 30–31
Rostow, W. W., 5, 8–9, 13, 14, 20
Russia: changing policies in, 9–10; and economic growth, 15, 16; and economic versus military power, 12; science and engineering graduates in, 8; space initiatives in, 17; and Toshiba Company, 79; and world arena of power, 9

S

Salt Lake City area, and entrepreneurial activity, 28
San Francisco Bay Area, and entrepreneurial activity, 28
Seattle: Puget Sound region of, 28; research park design at, 34
Sematech, 75–76
Smith, K., 26
Smith-Lever Act. See Agricultural Extension Act
Smithsonian Institution, 31
Society of University Patent Administrators (SUPA), 29
South Carolina, and vocational-technical education, 27
Southern Illinois University, Edwardsville, Center for Advanced Manufacturing, Production and Technology Commercialization at, 64
Soviet Union. See Russia
Sputnik, 8, 13, 22
Squibb Corporation, discovery of wonder drugs at, 40

Stanford Center for Integrated Systems, and microelectronic firms, 42
Stanford University: and corporations, 42; technology transfer offices at, 30
State Department, 5
State and local governments: and econometrics, 53–54; and land-grant universities, 55; and specialized university programs, 50–53; and university funding strategies, 55–59; and university legal advice, 54; and university research programs, 49–59; and university technical training, 52–53
Supreme Court, 25

T

Taiwan, and smokestack industries, 10
Tax Reform Act of 1986, 27
Technology diffusion: and the American research university, 5–20; current trends in, 10–13; retrospective view of, 6–10; university of the next century and, 17–20
Technology transfer: federal factors in, 23–25; new frontier of, 21–35; in the next decade, 34–35; origins of, 21–23; selected activity indicators in, 24–25; trends in, 29–34; and university research and development, 25–29
Texas: and university/corporate partnership, 75–76; and vocational-technical education, 27
Thomas Edison Program. See Ohio

U

United Kingdom: and diversification, 9; and higher education, 8; and new research projects, 75
U.S. Constitution, interpretation of, 13
U.S. General Accounting Office, 35
U.S. Patent and Trademark Office, 27, 32
University of Chicago, 7
University of Colorado, 12
University of Miami, and research grants, 80–81

University of Michigan, 14, 80
University of North Carolina, 28
University research administration: and corporate readaptation, 37-48; the future of, 73-85; and global technology diffusion, 5-20; and higher education, 37-48; and modern research centers, 61-72; new frontier of, 21-35; and university research programs, 49-59
University Research Centers, 61-72; activities of, 63-65; and Chambers of Commerce, 67-68; description of, 61-62; and employee relations, 68-70; financing of, 66-68; future of, 71; guidelines, 70-71; and motivation, 68-69; problems of, 62-63; rewards of, 70; staffing of, 65-66; working environment at, 69-70
University research and development: and economic development factors, 27, 28; and federal programs and tax incentives, 27; and higher education centers, 28-29; important factors for, 25-29; and industrial relations, 26; and international markets, 26; and new technology, 26
University of Texas, 76
University of Washington, 24-28, 31-32; Center for Process Analytical Chemistry (CPAC) at, 26-27; National Science Foundation (NSF) at, 26-27; Office of Technology Transfer at, 30; research park design at, 34; technology transfer recommendations at, 24-25; Washington Technology Center (WTC) at, 26-27
University of Wisconsin, 31
University/corporate partnerships: goals of, 45-47; in Texas, 75-76

W

Walcott, C. D., 31
Walt Disney: and adult audiences, 40; and Touchstone Pictures, 40
Washington, 28
"Washington First" policy, 28
Washington Research Foundation (WRF), 28, 31-32
Washington State legislature, 26
Washington State University, 32
Washington Technology Center (WTC), 26, 28, 32
Western Washington University, visual communications education unit at, 64
Whipple, R. P., 25, 35
Wilson, J., 46, 48
Wisconsin Alumni Research Foundation (WARF), 31-33
World Bank, 8
World War I, 7
World War II, 7, 26